T0187686

Holographic Interferometry

A Mach–Zehnder Approach

Holographic Interferometry

A Mach–Zehnder Approach

Gregory R. Toker

CRC Press
Taylor & Francis Group
Boca Raton London New York

CRC Press is an imprint of the
Taylor & Francis Group, an **informa** business

CRC Press
Taylor & Francis Group
6000 Broken Sound Parkway NW, Suite 300
Boca Raton, FL 33487-2742

First issued in paperback 2019

© 2012 by Taylor & Francis Group, LLC
CRC Press is an imprint of Taylor & Francis Group, an Informa business

No claim to original U.S. Government works

ISBN-13: 978-1-4398-8187-3 (hbk)
ISBN-13: 978-0-367-38164-6 (pbk)

This book contains information obtained from authentic and highly regarded sources. Reasonable efforts have been made to publish reliable data and information, but the author and publisher cannot assume responsibility for the validity of all materials or the consequences of their use. The authors and publishers have attempted to trace the copyright holders of all material reproduced in this publication and apologize to copyright holders if permission to publish in this form has not been obtained. If any copyright material has not been acknowledged please write and let us know so we may rectify in any future reprint.

Except as permitted under U.S. Copyright Law, no part of this book may be reprinted, reproduced, transmitted, or utilized in any form by any electronic, mechanical, or other means, now known or hereafter invented, including photocopying, microfilming, and recording, or in any information storage or retrieval system, without written permission from the publishers.

For permission to photocopy or use material electronically from this work, please access www.copyright.com (http://www.copyright.com/) or contact the Copyright Clearance Center, Inc. (CCC), 222 Rosewood Drive, Danvers, MA 01923, 978-750-8400. CCC is a not-for-profit organization that provides licenses and registration for a variety of users. For organizations that have been granted a photocopy license by the CCC, a separate system of payment has been arranged.

Trademark Notice: Product or corporate names may be trademarks or registered trademarks, and are used only for identification and explanation without intent to infringe.

Library of Congress Cataloging-in-Publication Data

Toker, Gregory R.
   Holographic interferometry : a Mach-Zehnder approach / Gregory R. Toker.
      p. cm.
   "A CRC title."
   Includes bibliographical references and index.
   ISBN 978-1-4398-8187-3 (hardback : alk. paper)
   1. Holographic interferometry. I. Title.

TA1555.T65 2012
543'.59--dc23
                                                    2011049568

Visit the Taylor & Francis Web site at
http://www.taylorandfrancis.com

and the CRC Press Web site at
http://www.crcpress.com

Contents

Section II   Mach–Zehnder Optical Shearing Holographic Interferometry of Compressible Flows

Section III   Mach–Zehnder Digital Holographic Interferometry and Related Techniques: Recording Mach–Zehnder Digital Interferograms/Holograms on CCD/ CMOS Sensors and Their Applications

Preface

This book is an original source of information on experimental methods of optical and digital Mach–Zehnder holographic interferometry, the approach which uses, for recording optical and digital holograms, standard photographic films and CCD/CMOS sensors having low resolving powers. The digital recording holograms on electronic CCD/CMOS sensors and their subsequent processing automatically falls under the category of Mach–Zehnder holographic interferometry because of the fact that the majority of CCD/CMOS sensors have low resolving powers ($<$50–100 mm^{-1}), which is comparable to standard photographic films. On the contrary, ordinary optical holograms are recorded on high-resolving-power photo materials ($\sim$500–10,000 mm^{-1}), so-called holographic photo materials.

This book provides an analysis of the problems of holographic registration of diagnostic waves, which carry not only useful phase information on the transparent objects under test, but also phase information on optical aberrations of Mach–Zehnder interferometers. The methods of visualizing a transparent object and measuring the refractive index of it (or its differentials) are also studied. The principles and methods of recording high quality reconstructed optical interferograms are described in detail, including the special coherent techniques applicable for generation and studying a signal wave and the technique of regulating the sensitivity of interference measurements.

The presented material is restricted mainly to studying transparent objects using methods of Mach–Zehnder classic, holographic reference beam, and shearing interferometry. Methods for mathematical and numerical processing of reconstructed interferograms for the purpose of retrieving phase data on transparent objects are discussed briefly, and only in an illustrative format, because these methods have already been analyzed accurately in numerous books on ordinary optical classic and holographic interferometry.

The problems of optical registration of Mach–Zehnder *reference beam holograms* and classic interferograms of transparent objects on standard photographic films are discussed in Chapter 1. Focused images of the transparent objects under study are recorded on photographic films having a low resolving power not exceeding $\sim$100 mm^{-1}.

The problems of reconstruction of focused image optical Mach–Zehnder holograms and studying signal waves are discussed in relation to the methods of optical filtration mandatory in the case of low spatial carrier frequencies of holographic patterns. Signal waves being optically reconstructed from Mach–Zehnder holograms are investigated by coherent interferometric techniques: two exposures, dual hologram, and real time. It has been shown that the signal waves can also be successfully studied by the additional coherent methods: holographic Schlieren and shadowgraph, Moiré deflectometry, and reference beam dual hologram shearing interferometry, including the analysis of the technique of regulating the sensitivity of interference measurements. Examples of studying transparent objects such as a supersonic air micro jet; vaporizing droplets of volatile liquids; thermal, acoustic, and shock waves in condensed matter; and optical breakdown in liquids are presented and discussed.

Methods of Mach–Zehnder *shearing holographic interferometry* and their application to aerospace engineering problems have been discussed very little in previous literature. The technique is analyzed in Chapter 2 in detail. Real holographic experiments on studying compressible shock flows in a supersonic wind tunnel and some other aerophysics objects are demonstrated and analyzed.

A series of existing interferometers intended for measuring compressible flow fields in wide-aperture gas-dynamic facilities are discussed in detail. Specific features of the proposed diffraction holographic shearing interferometer, which is well suited for the registration of phase information on compressible shock flows in a supersonic wind/shock tunnel, are analyzed. The presented scheme of reconstructing Mach–Zehnder shearing holograms and studying the process of interference signal and comparison reconstructed waves are discussed. A reconstructed shearing interferogram is free of aberration of the optical scheme of recording.

Studies of shock flows over sharp and blunt-nose aerodynamic models, spray jets, and turbulent and laminar convection streams as examples of applications, including methods of regulating the sensitivity of interference measurements, are performed and analyzed. The method of generating real-time shearing interference patterns is discussed.

The epoch of photographic films is coming to an end. The fast progress of the electronic industry offers scientists and engineers numerous modern types of electronic digital cameras having different temporal and resolution characteristics. In Chapter 3, algorithms for the *digital recording of* Mach–Zehnder holograms/interferograms on CCD/CMOS sensors are discussed. The main characteristics of the sensors responsible for the digital recording of transparent objects are analyzed. Attention is paid to the problems of optical imaging of Mach–Zehnder digital holograms on the sensors of CCD/CMOS cameras and the proper locations of CCD/CMOS camera lenses in schemes of holographic recording. The specific features of reference beam and shearing digital holographic interferometry are analyzed separately. Less attention is given to digital retrieving interference and holographic phase data, because numerous publications already have looked at the methods of retrieving phase information.

Related methods of optical diagnostics by using white light and coherent sources of radiation supplement the material stated in Chapter 3. Ordinary Schlieren and Moiré deflectometry techniques are analyzed in detail. Examples of applications for diagnostics of supersonic aerodynamic flow fields and subsonic hot turbulent jets are presented. Methods of spatial filtering Moiré deflectometry setup are analyzed. Separately, methods of optical filtering of diagnostic pencil-like beams of CW lasers and spatial filtering of collimated beams in the schemes of registration and reconstruction of Mach–Zehnder holograms are illustrated. Optical components and continuous as well as pulsed diagnostic laser systems, applicable for the holographic experiments, are discussed.

The author hopes that studying the presented materials will help research scientists and engineers using or planning to use methods of Mach–Zehnder holographic interferometry in their experimental projects in the field of aerospace engineering and noncoherent interaction of a laser light with matter.

All the information, well known as well as presented for the first time, is discussed on the foundations optical engineering and is suitable for graduate students of corresponding specialties. Numerous new and preliminary published designs of optical classic and holographic interference schemes of Mach–Zehnder help explain the main advantages of the proposed approach. Three dozen reconstructed and classic interferograms, deflecto- and Schlierengrams illustrate the presented material.

The author, a student of the Moscow Institute of Physics and Technology from 1968 to 1974, devotes this manuscript to former and present students of MIPT.

Gregory R. Toker
Research professor (KAMEA program) of the Technion
Haifa, Israel
2011
e-mail: gtoker@tx.technion.ac.il

Introduction

We must measure what is measurable and make measurable what cannot be measured.

Galileo Galilei (1610)

Mach–Zehnder holographic interferometry approach is a part of holographic interferometry, the science which describes methods of visualizing and measuring different transparent objects by advanced interference techniques. The distinguishing feature of the approach in comparison with ordinary optical holographic interferometry is that, for the purpose of recording holograms on photosensitive carriers, it uses materials and devices having a low resolving power. In this book, only standard photographic films and CCD/CMOS sensors will be discussed as the carriers of photosensitivity. It should be emphasized that, despite low resolving powers, the Mach–Zehnder holographic approach demonstrates all the fundamental opportunities characteristic of ordinary optical holographic interferometry. Analysis of those opportunities will be done in the first and second chapters.

First of all, the location of Mach–Zehnder holographic interferometry in the system of diagnostic techniques of transparent objects is outlined. Among different physical objects under study, there exists a very interesting and challenging class of *transparent objects*. Laser physicists and aerospace engineers encounter them during their everyday routine experimental practice. In aerodynamic applications, such objects are compressible flow fields and shock flows, laminar and turbulent convection flows, jets, and droplet sprays. In noncoherent laser–matter interactions, they include shocks; thermal and acoustic waves in solids, liquids, and gases; different types of plasmas arising due to the processes of optical breakdowns in solids, liquids, and gases; laser interactions with a solid surface in vacuum and atmosphere of gases, etc.

The mutual essential feature of such objects is transparency in the visible (and possibly in UV and IR) range. Transparency by itself creates additional obstacles for studying transparent objects in comparison with nontransparent ones from the point of view of photographing the objects. On the other hand, transparency creates favorable conditions for testing such objects in the optical range. The correct characterization of the transparent object under study is impossible without measuring its determinative parameter—the refractive index—which can be done in frames of *optical diagnostic tech-*

niques only. Being transparent, phase objects change the *phase* of a *diagnostic wave*, which penetrates through them, but not its *amplitude*.

There are five main visualizing and measuring optical diagnostic techniques: Schlieren,[1] shadow,[1] interferometry[2-4] (reference and shearing), and Moiré deflectometry.[5] All the optical diagnostic techniques measure *the refractive index or its derivatives*. Shearing interferometry, Moiré deflectometry, and Schlieren technique are similar to each other in some respects because they measure *gradients* of the refractive index. The Schlieren technique measures gradients in the direction perpendicular to an optical knife-edge (or other optical diaphragm). Shearing interferometry and Moiré deflectometry measure fringe shifts perpendicular to the correspondent fringes. Shadowgraphs visualize second derivatives of a transparent object, and finally, reference beam interferometry visualizes the refractive index.

The principal task of testing any transparent object by applying the diagnostic techniques is studying phase changes of the diagnostic wave. The final purpose is retrieving some experimentally achievable data on the refractive index of an object (or its gradients) and possibly 3-D mapping of it. The technical algorithm, for all visualizing techniques, is transferring phase changes in easily readable and measurable *amplitude changes* by means of imaging transparent objects on a photographic film or on the sensitive area of a charge coupled device (CCD)/complementary metal oxide semiconductor (CMOS) sensor.

If a transparent object refracts diagnostic rays at insignificant angles, then it is called a *phase object*. Insignificant angles are interpreted by different authors to be between 0.01 and 0.05 radians. In the presented monograph only phase objects and methods of recording and reconstructing their holograms will be analyzed and discussed.

The numerical and most versatile technique in the area of visualization and studying phase objects is *interferometry*. From the optical point of view, interferometry uses the phenomenon of interference of two coherent waves and thus converts changes of the phase in measurable changes of light intensity of the correspondent *plane amplitude interference pattern*. The simplest types of interferometry are the classic *two-beam reference*[2] and *shearing*[3,4] techniques.

The implementation of reference beam interferometry for the purposes of investigation of phase objects is based on splitting a *diagnostic beam* into two beams. The first of these, the *object beam*, is directed to propagate through a phase object; the second one escapes the object and stays the beam of *reference*. Recombination of these two beams at the output of an optical system, which is called in this case the reference beam interferometer, generates the appropriate interference pattern.

In shearing classic interferometry a collimated diagnostic beam is forced to propagate through a phase object. After that, the beam is divided into two identical waves, which are sheared relative to each other and appear as recombining waves, in order to record the shearing interferogram.

Good quality interference pattern, being the final product of the two interference techniques *with visualization of field*, is necessary for the successful procedure of retrieving the interference data for the purpose of obtaining the spatial distribution of changes of the refractive index or its gradients. It should be noted that methods of *amplitude division* of a diagnostic wave are more preferable in comparison with methods of the *front division*. This is due to the transverse structure of a diagnostic beam, which must be more uniform to be divided by the front division technique.

The simplest case is realized if an object has the flat form, i.e., the refractive index does not change in the direction of the object wave, propagating as a rule along the z-axis. The object's refractive index in this case is obviously a function of the two coordinates: $n = n (x, y)$. A more complicated situation arises if an object has the axis of symmetry. In this case, the optical diagnostics of a phase object should be realized in the direction which is perpendicular to the axis of symmetry. If gradients of the refractive index along the axis of symmetry, for instance x, are small enough, then for each element of the thickness of Δx, the refractive index can be examined as a function of the coordinates y and z only. In most experimental situations axial gradients are usually smaller than radial ones, or at least comparable. It is convenient to represent the refractive index as a function of two cylindrical coordinates: x and the radius $r = (y^2 + z^2)^{1/2}$.

If the refractive index of an object depends on all three coordinates, $n = n(x, y, z)$, retrieving the interference data becomes a complex tomographic task.[6] In this case, the spatial distribution of the refractive index can be evaluated only after retrieving data from interferograms obtained for a few different viewing directions of the probing beam.

The main purpose of interference measurements is to get numerical data on the 3-D value of the refractive index (or its gradients). The quality of an interference amplitude pattern is a necessary intermediate step. This purpose is achieved in classic interference techniques in one step, by recombining the object and reference beams (or two identical sheared beams in the case of shearing interferometry) at the output of the interferometer; in addition, the beams exist *simultaneously* in correspondent arms.

The idea of *optical holographic interferometry*,[7–9] unlike its classic counterpart, is based on a two-step approach. The first step is a preliminary encoding of the two object waves, signal, and comparison on a photographic film (photographic films) by means of the two *object holograms*. The object wave passing through the phase object is called the *signal wave*; the obtained hologram is also called the *signal*. The object wave in the case of absence of a phase object is called a *comparison wave*; the correspondent hologram is also called the *comparison*. Although a signal hologram carries useful phase information on an object and optical aberrations, the comparison hologram is recorded at the same conditions of the experiment as the signal hologram, but without the phase object under test. Consequently the comparison hologram carries only phase information on optical aberrations relevant to the setup of recording.

The second step is optical reconstruction of the signal and comparison waves from the signal and comparison holograms, and subsequent interference comparison of them in order to generate the output interference pattern. The comparison wave plays the same role as the reference wave does in classic interferometry—it plays the role of the wave of reference. Because the signal and comparison waves propagate the same optical path of an interferometer, they carry the same optical aberrations. At the output of a scheme of reconstruction, the interference of those two waves (optical subtraction) cancels aberrations; the resulting interference field is practically free of aberration of the interferometer.

At first glance this circumstance is not very important, especially for the phase objects responsible for a large interference fringe shift (thick phase objects). It plays a principal role in the case of thin objects, when a little fringe shift, of the order of 0.1, appears in the reconstructed interferograms. It will be shown that the process of measuring thin phase objects is possible only in frames of holographic approach, which guarantees the canceling of optical aberrations.

The principal advantage of holographic interferometry over its classic counterpart arises from the fact that a comparison wave propagates along the same optical path as a signal wave does (in the object arm); it exists when the object does not appear yet, i.e., before the run of a facility, or after. Signal and comparison holograms, unlike classic interferometry, are recorded at *different moments of time*. Thus, for the first step, the concept of *interference* of the two coherent waves, the object and comparison (or the two identical waves in the case of shearing interferometry), is implemented. The products of those interferences are two holograms: the signal and comparison.

For the second step, the recorded (signal and comparison) waves are reconstructed, due to the phenomenon of *diffraction*, from amplitude (or phase) holograms as optical gratings. It should be noted that an optical reconstruction of signal and comparison waves is more convenient to perform in a separate optical scheme.

As to *digital holograms* recorded on the sensitive areas of CCD/CMOS sensors, they could be automatically related to Mach–Zehnder type, because most sensors possess low resolving powers (<50–100 mm^{-1}). In digital holographic Mach–Zehnder interferometry, the resulting interference pattern (a digital reconstructed interferogram) is the product of a numerical computer treatment of the digital signal and comparison holograms recorded on CCD/CMOS sensors.

Ideally, optical holographic interferometry could be divided into the two following subdivisions (see Figure I.1): high and low carrier frequency holograms. The first technique requires a high-resolving-power ($\sim$500–10,000 lines/mm) photo material. As to the second one, it can be restricted by standard photographic films with resolving powers of <60–100 mm^{-1} and CCD/CMOS sensors having the same order values of resolving powers.

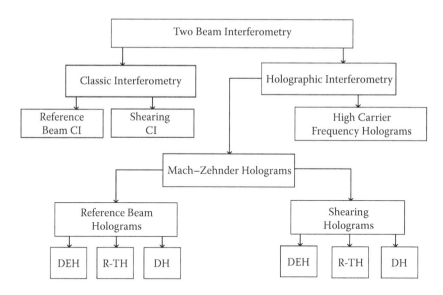

FIGURE I.1

Two-beam interferometry diagram. Abbreviations: CI—classic interferometry, DEH—double exposure holograms, R-TH—real-time holograms, DH—dual holograms.

Only the type of hologram called Mach–Zehnder is presented and analyzed in this book. It should be emphasized that Mach–Zehnder holograms possess all the advantages characteristic of holograms recorded on high-resolving-power photo materials. There are only two peculiarities: (1) lower spatial resolution and (2) mandatory requirements to reconstruct Mach–Zehnder holograms by collimated laser beam using methods of spatial filtering. It is interesting to remark that an object wave recorded in an optical scheme of a Mach–Zehnder interferometer by a pulsed laser may be reconstructed and studied by using a continuous wave (CW) laser in a scheme of reconstruction.

Let us sum up the principles of reference beam and shearing Mach–Zehnder holographic interferometry. A signal wave carries not only useful phase information on a phase object, but also phase information on optical aberrations of a Mach–Zehnder interferometer. The signal wave is encoded by means of the signal holograms on a low-resolving photographic film. The comparison wave carries phase information exclusively on optical aberrations of the interferometer. In a scheme of reconstruction, the signal and comparison wave are decoded from the corresponding holograms and interfere at the output. Because the object waves carry the same phase information on optical aberrations, the reconstructed interferogram is free of aberrations due to the process of optical subtraction (interference). Thus, the reconstructed interferograms carry only the useful phase information on a phase object.

In shearing Mach–Zehnder holographic interferometry the useful phase information on a phase object and aberration information on the scheme of recording are coded in signal and comparison shearing holograms. Decoded from the signal and comparison holograms, signal and comparison waves interfere, generating the reconstructed shearing interferograms without optical aberrations.

In classic optical interferometry, the reasonable requirement for permissible distortions of the resulting interference pattern is that they should not exceed $\lambda/10$ or $1/10$ of an interference fringe. This value correlates with the so-called visibly unresolved fringe shift, which as a rule is taken as ~0.1. In order to be measured, the minimal useful fringe shift in that case should exceed this value. In reality, the dynamic range of interference measurement is of about two to three orders of magnitude: from 0.1 to 10–20 fringes. Smaller fringe shifts cannot be resolved visually. Measurements of larger fringe shifts are restricted by the local spatial resolution of the technique. Thus, in classic interferometry it is impossible to measure useful phase information if the correspondent fringe shifts are smaller than (or equal to) the visibly unresolved fringe shift.

On the other hand, in the holographic technique, the visually unresolved fringe shift is fully determined by characteristics of the phase object under study but not by the distortions. By virtue of the *differential* character of the holographic technique, output distortions in the reconstructed interferograms are canceled, while in the classic technique they generally cannot be. The visually unresolved fringe shift may potentially be measured by applying methods to increase the sensitivity of optical interference measurements that will be shown below. The process of the CCD/CMOS recording holograms for measuring visually unresolved fringe shifts requires special analysis.

Owing to the differential character of the resulting interference picture, even complicated holographic interference schemes easily meet requirements for output permissible distortions $\leq 1/10$ of a fringe shift. The development and design of complicated classic interference schemes with the same level of distortions is not an easy task.

The *spatial carrier frequency* of a hologram depends upon the *angle of holographing* (i.e., upon the angle between the interfering object and reference waves) and the wavelength of the laser light. It is a very important feature if a hologram has either low or high spatial carrier frequencies, because methods of optical reconstruction depend on it; moreover the consequent holographic interference schemes are designed in different ways. If a hologram has a low spatial carrier frequency of *holographic fringes*, it is called a *Mach–Zehnder hologram*. This manuscript is restricted to the study of Mach–Zehnder holograms. As to holograms recorded with high spatial carrier frequencies, see References [7–10].

The selection of the applicable photo material also depends on the frequency of carrier holographic fringes. The essential parameter of any photographic material (photosensitive sensor) is the *resolving power*, which characterizes the maximum possible carrier frequency of interference pattern of a satisfactory quality recorded on it. Strict definition of the resolving power will be done in Chapter 1.

Registration of optical Mach–Zehnder holograms may be successfully realized on a standard photographic film having the resolving power of the order of ≤100 mm^{-1}. Owing to the fact that Mach–Zehnder holograms have low spatial frequencies, they should be reconstructed in optical schemes allowing *the procedure of spatial filtering* by illuminating the holograms with a collimated beam of a CW laser.

The holographic interference method can be implemented by means of three experimental coherent approaches (see Figure I.1): (1) real time, (2) double exposure, and (3) dual hologram techniques.

Real-time Mach–Zehnder holographic interferometry is the technique in which a signal wave existing in real time is forced to diffract on a *comparison hologram* obtained beforehand in the first order. The signal wave diffracted in the first diffraction order from the comparison hologram is compared interferometrically with the comparison wave, which is reconstructed by the reference wave from the same hologram and in the same direction.

In *double exposure* holographic interferometry, two holograms (a signal and comparison) exist during two consecutive laser exposures of an phase object by a pulsed laser (or two moments of time for a continuous wave laser) and are recorded on the same photo material. The resulting hologram is called a *double exposure hologram*. In this case the hologram is simply the superposition of the two holograms, the signal and the comparison. The signal and comparison waves are compared interferometrically after being reconstructed from the double exposure hologram.

In the *dual hologram technique* two waves, a signal (with an object) and comparison (without it), that exist during two consecutive exposures are recorded on two separate photographic films. One of them is the signal hologram; the other is the comparison. The two holograms, superposed in a *dual hologram holder*, are illuminated by the replica of reference wave in a scheme of reconstruction. The holograms generate two waves in the same order, the signal and comparison, which interfere and create the interference pattern necessary for retrieving phase data.

The advanced method of Mach–Zehnder holographic interferometry, where a signal wave, carrying phase information on the object under test, and the comparison wave are encoded on the corresponding holograms and then decoded for the purpose of retrieving phase data. This has the following two main advantages in comparison with its classic interference counterpart:

- First, the waves can be studied subsequently (on a specialized reconstruction scheme) by a CW laser.

- Second, the optical elements, which are used in holographic interference experiments, could be of a relatively low optical quality
  (Schlieren quality).

The classic interference technique formulates hard requirements for the
number and quality of the interferometer's optical elements and restricts
some possibilities of studying phase objects interferometrically. The critical
issue in classic interferometry is that object and reference waves exist *simultaneously*, propagating along the *different optical paths* of an interferometer,
through the different optical elements, because they are located in the *different arms*. The latter circumstance means that the optical elements in the
reference and object arms must be practically identical and consequently be
free from aberrations, creating output distortions. Such optical elements of
a high "interference" quality are usually expensive and hardly applicable to
some types of facilities. Furthermore, large-aperture optical elements could
be deformed under their own heavy weights and may introduce additional
noncompensated distortions as optical waves pass through them (in the
case of large-aperture interferometers). Although, in this case, experimenters replace large-aperture lenses with spherical mirrors having the same
wide working apertures and use other necessary tricks, the design of classic
interference schemes and their alignment is still a difficult, expensive, and
time-consuming task. On the other hand, the design of even a wide-aperture
holographic interferometer could be accomplished relatively quickly, easily,
and with small investments; in addition the set of the necessary optics is
more diverse.

From a practical point of view, the design of holographic interference
schemes is also very profitable. Holographic interferometers having apertures of any reasonable size can be designed on the basis of elements which
were developed for other optical schemes. For instance, a wide-aperture
shearing holographic interferometer can be constructed from optical elements that were intended for Schlieren devices in wind tunnels and other
powerful facilities, because the holographic technique allows operating with
optical elements having relatively low quality, i.e., "Schlieren" quality.

All these remarkable features of holographic interferometers are connected to the fact that two waves, i.e., the signal and comparison, pass
through *the same optical path*. They can be reconstructed and compared interferometrically after the run of a facility by using a specialized reconstruction
scheme. This master circumstance makes it possible to "soften" remarkably
the requirements in respect to the quality of the interferometer's optical elements. The same reason makes holographic interferometry simply one of
differential techniques. Indeed, if phase distortions in signal and comparison
waves are the same, they could be cancelled by the procedure of holographic
"subtraction" (interference), and fringe shifts in the resulting interference
pattern will depend only upon the phase changes of the object under test.

Generally speaking, the Mach–Zehnder holographic approach requires two specialized separate optical schemes. The first is simply a Mach–Zehnder interferometer for recording holograms, and the second is used for reconstructing the recorded wave/waves and its/their optical analysis. The process of reconstructing the signal wave allows studying of phase objects not only by interference, but also by applying a series of different diagnostic and visualizing techniques: Schlieren, shadow, and Moiré and reference beam dual hologram shearing interferometry.

Holographic recording possesses wide opportunities for studying phase objects by applying interference methods only. The reconstruction of object waves allows:

1. Acquiring a series of interference patterns with arbitrary direction and width of the background fringes having one signal hologram only

2. Enlarging or reducing the sensitivity of interference measurements

3. Performing interference comparison of two signal waves of the same object existing at different moments of time or different runs of the facility

4. Interference comparison of the two signal waves carrying information on different phase objects.

Classic and holographic interference schemes have the some accompanying problems: mechanical and thermal instabilities, problems connected to the polarization of a diagnostic beam, and quality of optical surfaces of optical elements.

Mechanical or acoustical disturbances could affect optical elements, forcing them to oscillate over stable locations in the scheme of an interferometer. The elements oscillating differently in reference and object arms of the interferometer could lead to random oscillations of the interference pattern during the exposure by a continuous wave laser. This effect can ruin the quality of the recorded interferogram/hologram. The exposure time of pulsed laser systems is very short, of the order of a few nanoseconds. The optical elements cannot change their positions during the laser pulse; nevertheless, mechanical instabilities may generate collateral fringes, deteriorating the useful system of fringes. There are a lot of possible sources of acoustical disturbances: a powerful facility itself during the run is a potential source of acoustical disturbances. Any powerful equipment in the neighborhood, including mechanical shutters, particularly the shutters of mechanical photographic cameras, are examples of potential sources of the acoustical noise. The result could be low quality of the recorded interferogram/signal hologram or even inability to record it at all.

Thermal instabilities are much slower than mechanical ones. They could be caused by air convection in the laboratory; for example, by a working device.

It is well known that two coherent optical waves may potentially interfere if they have the same plane of polarization. Numerous optical elements in an object and reference arms could change the plane of polarization of light at the output of the interferometer and worsen the quality (contrast) of the resulting interference pattern. This is owing to incoherent interference of the perpendicular components. The polarizing optical elements of an interferometer must be handled with particular care.

Scratches and dust particles on the surface of an optical element produce light scattering and lead to the losses of useful light. This effect is more observable in a narrow-pencil laser beam before its collimation. A poor or incorrect anti-reflection coating on the surface of an element also can lead to light reflection losses and to unwanted polarization.

All useful features and methods of classic and mainly Mach–Zehnder holographic techniques will be discussed below with demonstration of characteristics of the numerous phase objects under test. Holographic interference experiments presented in this book have been performed in the two areas of research: noncoherent laser–matter interaction and experimental aerodynamics. Some of the tested phase objects have been studied for the first time among other transparent objects, which have been visualized and investigated by the author during his 30-year experimental practice, primarily in Russian and Israeli laboratories.

Spatial features and principles of designing optical schemes for recording Mach–Zehnder holograms, as well as problems of mechanical stability of reference beam and shearing interferometers will be discussed. Characteristic features of the optical schemes of reconstruction, which analyze the front deformations of a signal wave, will be analyzed; coherent light sources for recording and reconstruction of plane-focused image phase and amplitude Mach–Zehnder holograms will also be considered. Some examples of retrieving interferometry data will also be discussed.

In this book, the phase objects are examined exclusively by two beam interferometry techniques. The author tried to focus attention on this holographic shearing interferometry technique,[11,12] which has not been discussed previously in any detail.

The author paid little attention in this text to the procedure of retrieving interferometric data, because it has been accurately discussed in References [3,4,7,10]. While writing this book the author was eager to apply an engineering approach to all stages of holographic interference experiments. He hopes that the material will be acceptable and easy to handle, not only for optics specialists, but also for laser and aerospace engineers, thermal physicists, plasma physics engineers, and for PhD and senior graduate students of corresponding specialties.

References

1. J. W. Beams, "Shadow and Schlieren Methods," in *Physical Measurements in Gas Dynamics and Combustion*, ed. R. W. Ladenburg, B. Lewis, R. N. Pease, and H. S. Taylor, Princeton University Press, Princeton, New Jersey, 1954.

2. R. Ladenburg and D. Bershader, "Interferometry," in *Physical Measurements in Gas Dynamics and Combustion*, eds. R. W. Ladenburg, B. Lewis, R. N. Pease, and H. S. Taylor, Princeton University Press, Princeton, New Jersey, 1954.

3. W. Merzkirch, *Flow Visualization*, Academic Press, London, 1987.

4. W. Merzkirch, "Generalized Analysis of Shearing Interferometers as Applied for Gas Dynamic Studies," *Applied Optics* 13, 409–413 (1974).

5. O. Kafri and I. Glatt, *The Physics of Moiré Metrology*, Wiley, New York, 1990.

6. D. Vukicevic et al., "Tomographic Reconstruction of the Temperature Distribution in a Convection Heat Flow Using Multidirectional and Holographic Interferometry," *Applied Optics* 28, 1508–1516 (1989).

7. R. Jones and C. Wykes, *Holographic and Speckle Interferometry*, Cambridge Studies in Modern Optics 6, Cambridge University Press, 1989.

8. P. K. Rastogi, ed., *Holographic Interferometry—Principles and Methods*, Springer-Verlag, Series in Optical Sciences No.68, 1994.

9. P. Smigielski, *in French, Holographic industrielle*, Teknea, Toulouse, 1994.

10. C. Vest, *Holographic Interferometry*, Wiley, New York, 1979.

11. A. Jones, M. Schwarz, and F. Weinberg, "Generalized Variable Shear Interferometry for the Study of Stationary and Moving Refractive Index Fields with the Use of Laser Light," *Proc. Roy. Soc. Lond.* A 322, 119–135 (1971).

12. G. Toker, D. Levin, and J. Stricker, "Dual Hologram Shearing Interference Technique for Wind Tunnel Flow Field Testing," *Experiments in Fluids* 23(4), 341–346 (1997).

About the Author

Gregory R. Toker was born in 1951 in Moscow, Russian Federation. In 1968, after 10 years of school, he entered the Moscow State Institute of Physics and Technology. He graduated from MIPT in 1974 and started to work at P.N. Lebedev Physical Institute of the Russian Academy of Sciences (RAS), later at the General Physics Institute of RAS—the two leading Russian Research Institutes in the field of physics and quantum electronics. His positions at these institutions included engineer, junior research scientist, and research scientist.

In 1992, he and his family emigrated from Russia to Israel, where he worked at the Technion—Israel Institute of Technology, Faculty of Aerospace Engineering (Haifa). In 1998, the family moved to Canada, where Dr. Toker worked for a few high-tech companies in Toronto and Ottawa. In 2005, he became a research professor at the Cuarnavaco University (Mexico). In 2007, Dr. Toker returned to Israel and presently works as a research professor in the Department of Physics at the Technion. He is married and has one son. He is a citizen of Israel and Canada.

Section I

Mach–Zehnder Optical Holographic Interferometry of Phase Objects

Introduction to Section I

In Section I, methods and special techniques of optical Mach–Zehnder holographic interferometry will be carefully presented. The content is restricted to an examination of reference beam Mach–Zehnder holograms.

All types of optical Mach–Zehnder holograms have been recorded, mainly on different types of standard photo films having ~100 mm^{-1} resolving powers.

Optical types of Mach–Zehnder holograms in some respects are ideologically close to each other. They have the same determinative characteristic—a low spatial frequency of signal and comparison holograms. In all cases the recording of optical holograms is restricted by the spatial frequency of ≤50 mm^{-1}. It should be noted that high-quality holographic gratings could be recorded at spatial frequencies in frames of ≈10 to 30 fringes per mm.

Different coherent methods of studying a signal reconstructed wave will be applied and analyzed, including double exposure, dual hologram, and real-time techniques. Some attention will be granted to the methods of holographic Schlieren, the shearing interferometry on the basis of identical reference beam holograms, and studying a reconstructed wave by Moiré interferometry (deflectometry).

It will be demonstrated experimentally that the imaging of fine phase details of the object under test are a function of the spatial frequency of a signal hologram.

The basic optical scheme for recording Mach–Zehnder holograms is the Mach–Zehnder classic interferometer. In Section I, it will be shown that Mach–Zehnder holograms differ practically from classic Mach–Zehnder interferograms only numerically. From the point of view of comparing classic and holographic approaches, a detailed analysis will start with studying the Mach–Zehnder interferograms. The main characteristics of photographic films for interferometry and holography, and principles of reconstruction of Mach–Zehnder optical holograms will be discussed in Section I. Techniques of enhancing the sensitivity of holographic measurements will also be covered. The letter technique will be analyzed from the point of view of measuring weak (thin) phase objects.

Numerous examples of applications of optical Mach–Zehnder holographic interferometry, mainly in the field of laser–matter interactions, will be also analyzed.

1

Mach–Zehnder Optical Reference Beam Interferometry

1.1 Using Mach–Zehnder Interferometers for Recording Focused Image Interferograms

The Mach–Zehnder interferometer was designed over 100 years ago[1,2] and up to now remains one of the most popular optical schemes for recording classic interferograms.

The standard optical scheme of a classic Mach–Zehnder interferometer, representing the division of amplitude interferometers, consists of two plane 100% reflection mirrors and two identical plane beam-splitting semitransparent mirrors. These four optical elements are located in the corners of a rhombus or a rectangle. It is convenient to replace the two semitransparent mirrors (or really optical wedges) with beam-splitting cubes as shown in Figure 1.1. The author used the presented scheme for recording interferograms of such phase objects as a supersonic air micro jets,[3,4] thermal waves in plastic,[5] optical breakdown of water[6] and alcohols,[7] etc.

1.1.1 The Optical Design of a Mach–Zehnder Interferometer

Figure 1.1 presents the classic Mach–Zehnder interferometer consisting of two flat reflecting mirrors $R = 100\%$, where R is the coefficient of reflection, and two identical beam-splitting cubes, which are used for the purposes of splitting and recombining object and reference beams at the output. Beam-splitting cubes are more preferable in comparison with semitransparent optical wedges, because of the simplicity of adjustment of the interferometer. In addition, beam-splitting cubes do not originate *optical path differences* between references and object waves in the middle of the beam splitting. The best beam-splitting cubes should have an antireflection coating to prevent ineligible reflections from the working faces.

It is worthwhile to analyze all optical elements of the interference scheme shown in Figure 1.1 to clearly understand their functions in the case of using a CW (1) as well as a pulsed laser system (17).

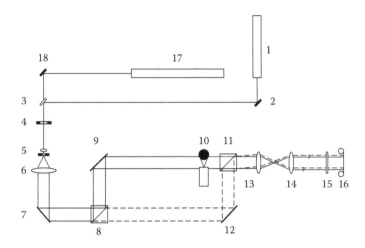

FIGURE 1.1

Focused image Mach–Zehnder interferometer for measuring phase objects. (1) CW laser; (2, 3, 7, 9, 12) flat Al 100% reflection mirrors; (4) diaphragm; (5) spatial filter; (6) collimating objective lens; (8, 11) beam-splitting cubes; (13, 14) telescope; (15) color filter; (16) photo film; (17) pulsed laser; (18) laser mirror.

If the CW laser works as the source of coherent radiation, its narrow beam illuminates a micro-lens of the *spatial filter* (5). The spatial filter consists of a micro-objective lens and a pinhole located at its back focal plane. At the output of the spatial filter a diverging spherical wave propagates free of optical aberrations and diffraction pictures caused by dust particles and tiny scratches on the front surface of the mirrors (2, 3). If the pinhole is located accurately in the front focal plane of the collimating objective lens (6), the parallel beam reflecting from the mirror (7) illuminates the input beam-splitting cube (8). To achieve this, the collimating lens must be carefully adjusted. The aperture of the lens (6) should be large enough to fill the aperture of the first beam-splitting cube (8), as the size of the beam-splitting cubes (8, 11) determines the field of view, which should be larger than the size of the phase object under investigation. The flat mirror (7) serves to turn the beam at 90 deg and to conserve space. Compact optical schemes are more stable mechanically. In further discussion it will be shown that reference beam interferometers are not tolerant to acoustical disturbances.

The first beam-splitting cube divides the collimated diagnostic beam into object and reference beams. It is assumed that both cubes have identical sizes and reflect and transmit the same portion of incident light. Only in this case will the output interference picture have the best possible contrast. The object beam is reflected by means of the diagonal beam-splitting plane with a partially reflecting coating. The beam is then reflected by a flat mirror (9) and probes the phase object under test. Passing through the second beam-splitting cube (11), the object wave is normally incident on a telescope (13,14) and then is recorded on the photographic film (16). The telescope serves for optical conjugation of a transparent object and the photographic film.

The reference beam, illustrated by the dashed lines, after the first beam-splitting cube, initially is reflected by a flat mirror (12) and then by the beam-splitting plane of the second cube (11), which serves to recombine the two beams at the output. If the back focal plane of the objective lens (13) coincides with the forward focal plane of the lens (14), then the incident beam and the reference beam at the output of the collimator are parallel where the ratio of the beams' diameters is proportional to the focal lengths of the lenses. The output diameter should fill the aperture of the photographic film (for standard 35 mm photo films, its diameter should be ≤25 mm).

A color band-pass filter (15) with a large transmission coefficient at the laser wavelength prevents the photosensitive material from being exposed to any collateral laboratory illumination. In the cases of highly luminous phase objects, narrow band pass or even laser line filters are required.

In the case of using the pulsed laser system (17) as a coherent light source, the CW laser (1) serves as the pilot laser. The Al-coated mirror (3) is replaced by a parallel glass plate (or an optical wedge) transparent to the pulsed laser light. The spatial filter (5), which could be damaged by a focused pulsed laser beam, should be replaced by a negative lens that has a focal plane which is coincident with the forward focal plane of the objective lens (6). The diaphragm (4) and laser mirror (18) are necessary for the adjustment and spatial coincidence of the beams from the CW and pulsed lasers.

1.1.2   Interference of Signal and Reference Waves

The complex amplitude of a plane wave can be written in the form: $A(x, y, z) = a(x, y, z) \exp[i\Psi(x, y, z)]$, where a is a real amplitude of the light wave, and Ψ is its phase.[8,9] The expression for an object (signal) wave propagating in the positive direction of the z-axis and carrying useful phase information on an object and aberrations can be written as:

$$A_o = a_o \times \exp(i\Psi_o) = a_o \times \exp[i(\varepsilon - \mathbf{k}_o\mathbf{r} + \varphi_o)] \quad (1.1)$$

where r is the radius vector in some system of coordinates; $\mathbf{k}_o$ is the propagation vector of the object wave with amplitude $2\pi/\lambda$; and ε is the phase change due the phase object under test. φ_o is the phase distortion of the object wave at the output of an interferometer (aberrations), for instance, owing to the viewing windows of an evacuated chamber and/or due to other optical elements located in the object arm of the interferometer. It is assumed that an optical scheme consists of relatively high-quality optical elements and aberrations are not large.

The reference wave can be written in the same form: $A_R = a_R \times \exp(i\Psi_R)$, where $\Psi_R = \varphi_R - \mathbf{k}_R \times \mathbf{r}$. Here $\mathbf{r}$ is the radius vector in some system of coordinates; $\mathbf{k}_R$ is the propagation vector of the reference wave with amplitude

$2\pi/\lambda$; φ_R is the phase distortions of the reference wave at the output of an interferometer.

The intensity of the interference pattern $I(x, y)$ of the output of the interferometer can be written in the form:[8,9]

$$I(x, y) = (A_o + A_R)(A_o + A_R)^* = A_o \times A_o^* + A_o \times A_R^* + A_R \times A_o^* + A_R \times A_R^* = a_o^2 + a_R^2 + a_o a_R\{\exp[i(\Delta\Psi)] + a_o a_R \exp[-i(\Delta\Psi)]\} = a_o^2 + a_R^2 + 2a_o a_R \cos(\Delta\Psi) \quad (1.2)$$

where the symbol * means complex conjugation, and $\Delta\Psi = \Psi_o - \Psi_R$. Equation (1.2) shows that the intensity of the light at some point at the output of an interferometer is determined by intensities of the object a_o^2 and a_R^2 reference waves, plus the interference term $2a_o a_R\{\exp[i(\Delta\Psi)]+\exp[-i(\Delta\Psi)]\}$, which depends on the phase difference $\Delta\Psi = \Psi_o - \Psi_R$.

In order to successfully record the interferogram on the photographic film (16), the phase difference should not change remarkably during the exposure time. For short ~5- to 30-nanosecond pulses of a Q-switched laser (17), this condition is easily achievable. As to the exposure time of a CW laser system (1), it could be a complicated problem for long, ~1 millisecond and longer, exposures, because optical elements in the object and reference arms and the holder of a photo camera with the photographic film (16) oscillate independently under an influence of acoustic disturbances. For a long exposure, the value $\Delta\Psi = \Psi_o - \Psi_R$, which changes chaotically during the exposure, may dramatically diminish the contribution of the interference term. In the worst case, when a statistically averaged value of the cosine is equal to zero, the resulting intensity of the light in the interference pattern is composed from the sum of the intensities of the waves, which simply means the addition of incoherent light beams.

The magnitude of the interference term in Equation (1.2) determines the modulation of the light intensity, and ultimately, it is responsible for the quality of the interference picture. When the term is small enough, the phase information encoded in the interference pattern can be lost. From a practical point of view, this means degradation of the quality of the interferogram (a low contrast) to the point where the phase data cannot be reduced.

1.1.3 Characteristics of Interference Pattern

The phase difference $\Delta\Psi = \Psi_o - \Psi_R$ can be presented in the following form: $[\varepsilon - (\mathbf{k}_o - \mathbf{k}_R) \times \mathbf{r} + \varphi_u]$. The value of $\varphi_o - \varphi_R = \varphi_u$ illustrates uncompensated aberrations between the object and reference beams. The propagation and radius vectors in the system of coordinates connected to the process of recording are written as:

$$\mathbf{k}_o = [0, 0, k]; \mathbf{k}_R = [k \sin(\alpha), 0, k \cos(\alpha)]; \mathbf{r} = [x, y, z] \quad (1.3)$$

The incident object wave is normal to the plane of the photographic film: (x, y). As to the propagation vector of the reference wave, it should be noted that it is located in the (x, z) plane, and a is the incident angle of the reference wave (or the holographing angle or *offset* angle). Taking into account that, in the plane of the photo film $z = 0$, the phase difference can be written as $\Delta\phi = \varepsilon - kx\cdot\sin(a) + \varphi_u$. The second term can be presented in the form $2\pi fx$, where $f = \sin(a)/\lambda$ is the *spatial frequency* of a *background interference pattern*.

As it was noted in the introduction, the maximal allowable distortions at the output of a satisfactorily designed interferometer should not produce fringe shifts that exceed one tenth of λ; consequently, the phase distortions should be smaller than $2\pi/10$: $\varphi_u < 2\pi/10$. If the output phase distortions are neglected, phase modulation in the background interference pattern depends only on characteristics of the phase object under test. Amplitudes of interfering beams and Equation (1.2) can be rewritten as:

$$I(x,y) = a_o{}^2 + a_R{}^2 + a_o a_R \exp\{i\,[\varepsilon - (\mathbf{k}_o - \mathbf{k}_R) \times \mathbf{r}]\} + a_o a_R \exp\{-i[\varepsilon - (\mathbf{k}_o - \mathbf{k}_R) \times \mathbf{r}\}$$
$$= a_o{}^2 + a_R{}^2 + 2a_o a_R \cos[\varepsilon\,(x, y, z) - 2\pi\,fx] \qquad (1.4)$$

In the absence of the phase object the expression (1.4) can be rewritten in the form:

$$a_o{}^2 + a_R{}^2 + 2a_o a_R\cdot\cos[kx \sin (\alpha)] = a_o{}^2 + a_R{}^2 + 2a_o a_R\cdot\cos[2\pi \times f\,x] \qquad (1.5)$$

Equation (1.5) represents the expression for the background interference pattern. A simple analysis of Equation (1.5) shows that, in the x direction, the intensity of the light changes harmonically with the step or *spacing*:

$$d = \frac{\lambda}{\sin(\alpha)} = \frac{1}{f} \qquad (1.6)$$

In other words, the harmonically changing intensity of the light in the interference pattern in the x direction represents a system of bright and dark fringes with the spatial frequency f. The spacing d is called the *width of a fringe* and is defined as the distance between any two bright or dark fringes. This value is measured in mm^{-1} = lines $\times$ mm^{-1}. The expression "one line per millimeter" means that both dark and bright fringes fit in one millimeter.

It can be seen from Equation (1.5) that the phase in the background pattern changes in the x direction with a constant gradient: $const = d(2\pi fx)/dx = 2\pi\cdot\sin(a)/\lambda$. Thus the background interference field is generated with linear change of the phase; therefore, the sign and value of the offset angle α is determined by the sign and the magnitude of the background phase gradient, i.e., they depend upon the position of the reference beam.

Under these circumstances it is obvious that studying phase objects having the steeper spatial changes of the refractive index (phase) requires the background interference pattern having the higher spatial frequency. The same approach is applicable to small phase objects. From an experimental point of view, it means that for studying such phase objects, larger angles between the object and reference beams must be chosen. For example, if an object has the linear size of the order of ~1 mm, the object to "be resolved" should be "covered" with at least 10 interference fringes of the same color; thus the offset angle is calculated from the relation: $\alpha > \arcsine (f \times \lambda) \approx 0.36$ deg, $\lambda = 632.8$ nm and $f = 10$ mm^{-1}.

The emergence of a phase object in the object arm will lead to the phase change $\varepsilon(x, y, z)$ and, consequently, to the phase modulation of the correspondent interferometric picture. As a result, the light intensity in the imaging plane will be changed. The system of background fringes is no longer equidistant and straight; therefore, fringes in the area of the phase object are "shifted." Analysis of Equation (1.4) shows that any fringe represents the locus of a constant phase. In the case of the background interference picture, this it is simply a straight line.

If the effect of refraction of an object wave propagating through the phase, nonuniformity is negligible. The transparent object under study is called the "phase object," and the fringe shift $\Delta N(x, y)$ in the given point (x, y) of the interference field in the direction perpendicular to the system of undisturbed background fringes is proportional to the *optical path difference* expressed in wavelengths of the diagnostic light:[8,9]

$$\Delta N(x, y) = \frac{\varepsilon(x, y)}{2\pi} = \frac{1}{\lambda} \int \left[n(x, y, z) - n_0 \right] dz \qquad (1.7)$$

Here n_0 is the refractive index of the undisturbed medium, n is the index of refraction of the object under test and λ is the wavelength of the diagnostic light. If, for instance, $\varepsilon(x, y) = 2\pi$, the fringe shift is equal to one fringe and the optical path difference is one wavelength. Thus, phase changes in the object wave are converted into the changes of light intensity in the interference pattern, and the phase object is "visualized."

1.1.4 Infinite- and Finite-Width-Fringe Interferograms

The interference pattern described by Equation (1.4) is called the *finite-width-fringe interferogram*. If the offset angle α in Equation (1.6) is reduced, then the width of a fringe d increases, up to the point where it may exceed the size of the field of view (the size of the frame). In this case, the reference and object waves interfere in phase (or multiple 2π) intensifying each other, and the field of view is filled out with one bright fringe. This means that the interference pattern depends only upon characteristics of the phase object:

$$I(x, y) = a_o^2 + a_R^2 + 2a_o a_R \times \cos[\varepsilon (x, y)] \qquad (1.8)$$

Under these conditions the transition to the next bright (or dark) fringe means that the corresponding optical path difference is equal to one wavelength (one wavelength longer or shorter). Closed lines represent borders between bright and dark fringes. The interference pattern in the form of Equation (1.8) is called the *infinite-width-fringe interferogram*. Unlike a finite-width-fringe interferogram, the information about the sign of gradients of the refraction index in the corresponding infinite-width-fringe interferograms is lost, and any closed line represents the locus of a constant phase. Thus an infinite-width-fringe interferogram clearly visualizes distinctive boundaries of the phase object, which could be very useful in some applications. The ideal series of the interferograms narrowly describes some transparent object that consists of infinite-fringe and a few finite-fringe interferograms having different spatial frequencies and directions of the fringes.

1.1.5 Alignment of the Interferometer

By using the narrow pencil beam of a CW laser, a preliminary alignment of the interferometer's optical scheme could be performed. In Figure 1.1 the optical elements (5, 6), the phase object, and the collimator (13, 14) should be temporarily excluded from the scheme. At the first step of alignment the experimenter is sure of horizontal positions of the reference and objects beams, and all optical elements are centered with respect to the beams. The spatial coincidence of the reference and object beams must be achieved by means of the rotating mirrors (9, 12) of the interferometer and by aligning the beam-splitting cubes located on kinematic optical mounts or tables.

The spatial coincidence of the reference and object beams in a far field is confirmed by the coincidence of the system of circular rings due to the diffraction on the diaphragm (4). To visualize in a far field the interference picture in finite-width fringe, one may temporarily place a long-focal-length lens at the output of the interferometer. The spatial frequency of a carrier fringe pattern and orientation of the interference fringes are controlled by the rotating mirrors (9, 12).

The last step of alignment is the sequential repositioning of optical elements on their places. Special attention should be paid to the adjustment of the spatial filter (5) with simultaneous centering of the collimated diagnostic beam. The final adjustment of the interferometer in collimated beams is realized by changing the spatial frequency and orientation of the fringes in the background interference pattern at the output.

Photographic film (16) is located in the interference field perpendicular to the object beam on the z-axis. The plane of the photographic material must be optically conjugated with the middle cross section of the phase object under study in the z direction. Technically it can be done by placing an optical target (a wire, a needle) in the middle cross section of an object. The target is illuminated by a white light source, which has a very low spatial coherence that makes focusing easy and sharp.

1.2   Characteristics of 35-mm Photographic Films for Mach–Zehnder Interferometry

Practically any available 35-mm photographic film can be used for recording Mach–Zehnder interferograms. Imaging a phase object on the photographic film requires a mechanism for holding the film, changing a shutter speed, and advancing the film. For these purposes, the most pragmatic approach is to use a standard 35-mm photographic camera. Without the camera's lens, it represents the simplest type of interferometric camera. The film can be easily advanced to the next frame by means of a mechanical or motor drive. Cameras are equipped with a tripod socket, so that an appropriate post can be attached to the camera as to any other interferometer's optical element. The author has successfully used different types of mechanical photo cameras in numerous interference experiments. The only drawback of the mechanical systems is that their shutter speeds generate acoustical disturbances during the process of taking the pictures. In order to avoid this, one should produce an acoustical isolation between the camera and the rest of the interference scheme. The best choice is using a photographic camera with an electronic shutter, which does not generate acoustical disturbances.

The chemical industry produces numerous types of 35-mm perforated photo films, with panchromatic as well as orthochromatic spectral characteristics, which are applicable for interference measurements. Some essential features of these films deserve further discussion.

A photographic film consists of a photographic emulsion coated on a flexible transparent substrate (acetate film, for example) or on other flexible substructures. The photo emulsion of standard photo materials consists of a suspension of crystals of silver halides in colloid, a gelatin layer. The thickness of the emulsion varies in limits of a few microns. The thickness of photo films usually does not exceed the value of the order of 100 to 200 microns. The 35-mm film frame of the so-called "C" format has the aspect ratio 3:2 (width to height) with the maximal size of a rectangular picture of 36 × 24 mm.

The optical scheme of the Mach–Zehnder interferometer described in Section 1.1 is well suited for recording reference beam interferograms of phase objects which do not exceed the aperture of the beam-splitting cubes. Spatial resolution of the technique is dependant on the resolving power of the photo film (16) and the telescopic system (13, 14).

In the case of magnification of a phase object, the experimenter should try to minimize the distances between the pupil plane of the objective lens (13) and a phase object, because the minimal distance is determined by the linear size of the beam-splitting cube (11). This circumstance prevents using a micro-objective lens as the first lens (13) of the collimator for biggest magnification, because micro objectives have small focal lengths. The

coefficients of magnification are also limited by the approximate value of 8×
to 10×, because at higher magnifications in standard photo emulsions having resolving powers of the order of ≤100 lines × mm^{-1}, the individual silver
grains become distinguishable; the signal-to-noise ratio (SNR) decreases,
consequently reducing quality of the image. To avoid these effects the magnifications should not be excessive.

Transmission characteristics of a negative are expressed in the experimentally obtained dependence of the intensity transmittance coefficient T as a
function of the light intensity I, for which the sensitometric variables D and
H are used in the form of dependence of optical density D versus *log* exposure H:[8]

$$D = \gamma \log H \quad (1.9)$$

The value of γ is called the contrast coefficient. The optical density reflects
the approximate logarithmic sensitivity of the human eye to changes of a
light's brightness. The optical density D is defined as the logarithmic function of the coefficient of intensity transmittance T, and for a processed photo
film is expressed as:

$$D = \log\left(\frac{1}{T}\right) : T = \tau^2 = 10^{-D} \quad (1.10)$$

Exposure H is determined as the product of light intensity I and time
exposure t: $It = H$. Being displayed graphically, this relation is called the
Hurter–Driffield (H-D) curve, and can be found in Reference [8]. The curve
(H-D) can be divided into three regions: (1) little exposures (region of under-
exposures [$<H_1$]), (2) mediate exposures (normal exposures [$H_1 - H_2$]), and (3)
large exposures (region of overexposures [$>H_2$]). The slope of the curve in
the region of normal exposures, where the optical density changes approximately linearly, is determined from geometrical considerations (α_D is the
angle of the slope) as:

$$\gamma = \tan(\alpha_D) = \frac{D_2 - D_1}{\log\left(\dfrac{H_2}{H_1}\right)} \quad (1.11)$$

Generally speaking, the contrast is also determined as the range of difference from light to dark areas of a negative: $C = T_c/T_d$, where T_c and T_d are
the intensity transmittance for a clear and a dark fringe, respectively. When
photographic material has a large γ, it is called a "contrast" material, if $\gamma > 2.0$.
For contrast photographic materials the optical density of a negative changes
more rapidly with the exposure. The contrast types of photo materials are

more applicable for use in Mach–Zehnder interferometry and holography than standard films with medium values of γ (≈ 0.75). Contrast materials are associated with the better transmitting characteristics of interferograms/ holograms, and facilitate better decoding interferograms for the purpose of retrieving phase data. An exposed and processed photo film (negative), even in the linear range of optical densities, demonstrates nonlinear dependence on the intensity of light for the intensity coefficient of transmittance I:

$$T_2 = T_1 \left(\frac{I_2}{I_1} \right)^{-\gamma}$$

(1.12)

Equation (1.12) can be easily derived from expressions (1.10) and (1.11). This means that the correspondent interference grating (the negative) is not sinusoidal, because the transmittance does not change harmonically with the coordinate x, as in the case of the light intensity (see Equation [1.4]). It should be noted, however, that this circumstance does not affect the accuracy of fringe shift measurements. Fringe shifts on an interferogram are usually evaluated by means of the exact locations of the fringe's centers; thus the results of the reduction of interference data are hardly dependent on the characteristics of the coefficient of intensity transmittance. The parameter characterizing a modulation of the intensity of light, and consequently the quality of the interference picture, is called the visibility of fringes and is determined as follows:

$$V = \frac{I_{max} - I_{min}}{I_{max} + I_{min}}$$

(1.13)

where I_{max} is the intensity in the center of a bright fringe, and I_{min} is the light intensity in the center of a dark fringe. The visibility, broadly speaking, is a function of coherent characteristics of the laser light and the ratio of amplitudes of the object and reference waves a_o/a_R (see Equation [1.4]):

$$V = \frac{2 \dfrac{a_o}{a_R}}{1 + \left(\dfrac{a_o}{a_R} \right)^2}$$

(1.14)

If the object and reference waves have the same amplitudes, the minimal intensity $I_{min} \sim 0$, the maximum $I_{max} \sim 1$, and the visibility is maximal and close to the unity, $V \approx 1$. Here it is supposed that the interferogram is recorded with the laser light of a very large length of coherence. The interference picture with a low ratio of a_o/a_R is characterized with a low value of visibility; for instance, $a_o/a_R = 0.25$ $V \approx 0.47$. If this interference picture is focused on a photo material, the negative (recorded interferogram or hologram) probably will have a low contrast. In the case of $a_o/a_R = 1$, the negative has maximum possible contrast and modulation characteristics.

The substantial parameter characterizing modulation of the coefficient of intensity transmittance of a photo material is the modulation M, which is expressed as:

$$M = \frac{C-1}{C+1} = \frac{10^{\Delta D} - 1}{10^{\Delta D} + 1} = \frac{T_c - T_d}{T_c + T_d} ; \quad \Delta D = \log\left(\frac{T_c}{T_d}\right) \tag{1.15}$$

where C is the contrast of the photo material. The modulation of a photo material is a function of the contrast and changes between zero and unity.

The modulation of the coefficient of intensity transmittance of some hypothetical but ordinary photographic film may be easily evaluated. The exposure latitude of the most conventional films is $\log(H_2/H_1) \sim 1.5$, and the contrast coefficient is on the order of 0.75. The exposure latitude is determined as the range of exposures from underexposed to overexposed negatives that nevertheless produces a negative of an acceptable quality. From Equation (1.11) the interval of optical densities can be found as $\Delta D \sim 1.1$. Equation (1.15) gives us $M \approx 86\%$. It is useful for the experimenter if the value of fringe shifts can be evaluated accurately. Better contrast is achievable in the case of contrast photo materials, because they have larger maximum optical densities D_{max} and larger coefficients of contrast γ_{max}.

Table 1.1 shows some image structure characteristics for five standard panchromatic types of 35-mm photo films having an equal film speed of ISO (ASA) 100. Film speed is the sensitivity to light, as indicated by a number (100, 400, etc.). The higher the number, the more sensitive or faster is the film.

TABLE 1.1

35-mm Panchromatic Negative Photographic Films for Interference Measurements

Film/Speed	Resolution [Lines × mm^{-1}]	Spectral Sensitivity [nm]	γ_{max}/D_{max}
Agfa APX/100 by half at 60	150 (1000:1) <660	0.75/2.5	
Fuji Neopan Acros/100	200 (1000:1)	<640	0.75/2.25
Kodak T-max/100 by half at 120	180 (1000:1) <690	0.75/2.5	
Agfa Vista/100	130 (1000:1)	<620	0.75/2.5
Kodak Tri-X Pan/400	by half at 60	<630	0.75/2.5

All these photographic films and others with approximately the same standard value of contrast and the same level of resolution can be used for interference and holographic measurements. Kodak Tri-X Pan film, having a speed of 400 ASA (ISO), was put in this table, because it was one of the first photographic films to have been successfully used for recording low spatial carrier frequency holograms.[10]

The resolving power of a photo material is one of the essential characteristics, which plays a more important role in Mach–Zehnder holographic interferometry compared with its classic counterpart because of higher spatial frequencies of holograms. The resolving power characterizes the ability of a photographic film to show separation between closely spaced lines of some test chart. This parameter refers to the film's ability to distinguish fine details of the phase object under test and is measured in lines per millimeter. It was established experimentally that the resolving power of a photographic film is a function of the contrast, because the resolution is indicated for two types of test chart: a low-contrast chart with the luminance ratio of 1.6:1, and a high-contrast ratio chart with the ratio 1000:1. The resolving power of a photographic film could be evaluated due to the fact that the higher the spatial frequency of an interferometric pattern the lower the contrast of the negative.

It is well known that, due to the scattering a light in photo emulsion, the contrast of the photo film depends on the spatial frequency of interference fringes and drops when the frequency is getting higher. In accordance with Equation (1.15), this leads to worse modulation of the coefficient of intensity transmittance versus the spatial frequency of interference fringes. The resolving power of the negative also becomes worse.

The problem of specifying resolution as a function of the spatial frequency can be solved by the introduction of the Modulation Transfer Function (MTF). Physically, the MTF describes the results of light scattering in the emulsion dependant on the spatial frequency of an interference picture. In fact, MTF characterizes the spatial frequency response of a photo film and can be expressed by using the function of modulation: MTF = (modulation)/(modulation at a low spatial frequency). Thus the resolution of a photo material more reasonably can be characterized by the spatial frequency at which MTF is 50% (i.e., the modulation has dropped by half). These spatial frequencies are shown in Table 1.1 together with the resolving powers.

It is worth noting in conclusion, that contrary to its simplicity, focused image reference beam Mach–Zehnder interferometry is a flexible and at the same time powerful technique, which could be widely adopted in aerospace and laser applications laboratories, mainly for relatively small phase objects. The technique does not need special photo materials, because the Mach–Zehnder interferograms can be successfully recorded on standard photographic films, and the spatial resolution of the image is determined by the optical quality of a focusing objective lens and a photographic film. The quality of beam-splitting cubes in the scheme of a Mach–Zehnder interferometer is also responsible for resolution characteristics of recorded interferograms.

A conventional camera (without its lens) loaded with ordinary photo film could be successfully used for interference experiments. The only shortcoming is that the technique has the relatively low mechanical stability of the interference scheme of recording.

1.3 Applications of Mach–Zehnder Interferometry by Imaging on Photographic Films

1.3.1 Supersonic Micro Jet Impinging of a Spherical Target

The two focused-image interferograms presented in Figure 1.2 illustrate the application of Mach–Zehnder interferometry to visualizing and measuring a supersonic air micro jet impinging on a spherical target.[11] High-pressure air at 7 atm is released through a converging nozzle that is 1.3 mm in diameter. Interferograms were recorded on a contrast photo film with the resolving power of ≈ 200 mm^{-1}. The spatial resolution in the field of visualization, which is of the order of ~40 μm, is defined by the spatial resolution of a focusing lens and a relatively long distance between the lens and the air jet.

On the finite-width-fringe interferogram, the spatial frequency of the background fringe pattern is $f_y = 0.5$ mm^{-1}. The phase changes linearly in the

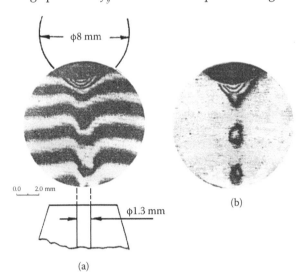

FIGURE 1.2

Mach–Zehnder interferograms of a supersonic air jet impinging on an 8-mm-diameter steel sphere: (a) finite-, (b) infinite-width-fringe interferogram. The laser wavelength of the diagnostic beam of an argon ion laser is 514.5 nm. The magnification is 1.42×. The size of the minimal resolvable phase nonuniformity is ~40 μm.

vertical direction along the micro jet with the constant gradient of $2\pi f_y = \pi$ radian $\times$ mm^{-1}. The offset angle is equal to $\alpha = \arcsine(\lambda \times f_y) = \arcsine(0.5145 \times 10^{-3} \times 0.5) = 0.26 \times 10^{-3}$ radians, that is, approximately 0.9 angular minute. Far away from the target, in a free stream of the jet, the fringe shift is about 0.6 fringe and the phase change in the free jet is about $0.6/2\pi$. According to Equation (1.7), the change of the refraction index can be calculated as: $\Delta n = \Delta N \lambda / D_j$, where ΔN is the fringe shift, and $D_j \approx 1.7$ mm is the averaged diameter of the jet. On the other hand, $\Delta n = K\Delta\rho$, where $K = 0.22 \times 10^3$ cm^3/g is the Gladstone–Dale constant for air, and $\Delta\rho$ is the change of air density. Hence, the change of air density is evaluated as $\Delta\rho \approx 0.82 \times 10^{-3}$ g $\times$ cm^{-3}.

An abrupt change in the air density can be observed in the vicinity of the target. The phase gradient in this region noticeably exceeds the gradient in the undisturbed region of the free micro jet. At the distance of one bright background fringe three interference fringes are observed, which means that air density enlarges up to $\Delta\rho \approx 4 \times 10^{-3}$ g $\times$ cm^{-3} in the vicinity of the target's surface. The analysis of the infinite-fringe interferogram confirms the above approximation.

1.3.2   Laser Drilling a Plastic

During the process of a laser beam interaction with biological hard tissues, the excessive thermal fluxes may lead to the necrosis of the living matter. In particular, high thermal fluxes accompany the laser drilling deep and narrow channels in hard biological tissues. The process of drilling was modeled by means of the interaction of CO_2 and Er:YAG focused laser beams with one of the popular transparent plastics (Plexiglas). The laser beams were focused on focal spots 0.45 (CO_2) and 0.5 (Er:YAG) mm in diameter on a 2-mm slab of the plastic.[5]

During the process of drilling the hot vapors of the plastic expanding in atmosphere efficiently transfer heat to the walls of a narrow laser channel, which might lead to thermal destruction of biological tissue over a great depth and cause extensive necrosis. In order to model[5] such overheating situations, CO_2 laser pulses were used with the following parameters: $\varepsilon = 100$ mJ, $\tau = 5$ μs. The intensity and energy density (flux density) were $q = 12.6$ MW $\times$ cm^{-2} and $E_s = 63$ J $\times$ cm^{-2}, respectively. The upper laser channel having the form of a cone with a ~0.5-mm base and the height of ≈6 mm is imaged in Figure 1.3(a). The channel was obtained by applying 100 laser pulses with the averaged capacity of 38 μg $\times$ J^{-1}.

The Mach–Zehnder interferometer was applied to visualize and measure the refractive index of the Plexiglas slab *in situ* in the region affected by a strong thermal wave propagated in the radial direction. One of the focused image interferograms is shown in Figure 1.3(b). One can see in the interference picture recorded long after propagation of the thermal wave that the laser channel is surrounded by a conical area of the plastic, which was strongly overheated, and resulted in an irreversible change of the refractive index $\Delta n \approx 2 \times 10^{-3}$.

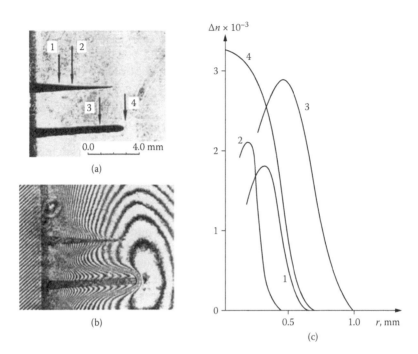

FIGURE 1.3
Laser drilling of 2-mm Plexiglas slab by radiation of CO_2 (λ = 10.6 μm) and Er:YAG (λ = 2.93 μm) laser pulses. (a) Focused image of laser channels in Plexiglas in the light of a He-Ne laser (λ = 632.8 nm). The upper channel was formed due to a CO_2 laser light. The lower channel is formed due to Er:YAG laser beam. (b) Interferogram of laser channels in the light of a He-Ne laser. (c) The graph calculated from the interferogram (b), showing radial distributions of refractive index in Plexiglas over channels created by the overheating thermal waves. Distances between focal planes and cross sections 1, 2, 3, and 4 are 2.4, 3.8, 5.7, and 7.1 mm, respectively.

Calculations of the radial distribution of the index of refraction $\Delta n(r)$ for cross sections 1 and 2 (see Figure 1.3[c]) indicate that the refractive index is not a monotonous function on the radius; it reaches maximum at a certain distance from the internal wall of the channel. The same behavior is also true for the refractive index in the channel, which is produced by focusing the radiation of an Er:YAG laser (λ = 2.94 μm) into a spot with a diameter of ≈0.5 mm by a BaF_2 lens with F = 200 mm. The lower channel having the length of ~7 mm was formed by 50 laser pulses (ε =1 J, τ = 250 μs) with an average capacity of about 32 μg × J^{-1}. In this case, the maximum change of Δn becomes greater than for CO_2 laser drilling values: 3 × 10^{-3}. Calculations were performed for cross sections 3 and 4. The results are due in part to more intense heat flows associated with the relatively low value of the absorption coefficient α_{ab}, at the wavelength λ = 2.94 μm (α_{ab} = 38 cm^{-1}).

Note that a remarkable nonflatness on the order of 6 to 12 wavelengths (λ = 632.8 nm) of the Plexiglas slab distorts the probing wave, leading to irregularity of the system of the interference background fringes. This is in contrast with the high quality of the background pattern in air (see Figure 1.3[b]) in the vicinity of air–plastic boundary. In Section 2.2 it will be shown that such aberrations can be cancelled by applying the holographic approach.

1.3.3   Laser Breakdown of Tap Water and Ethanol

Laser breakdown in liquids (particularly in tap water) is of considerable interest due to its potential application to online analysis of suspended solid contaminants.[6] Most studies of the phenomena of optical breakdown in water refer to the integral dynamics of the water heated by laser radiation, and only a few studies have been done regarding the structure of the region of breakdown. Taking into account that water (in general) and tap water (in particular) contains numerous suspended particulates, the laser breakdown events must have a discrete character. Therefore, studying only the integral parameters of water cannot adequately describe the process. It seems that a more detailed dynamic description of the plasma column is necessary, which can be realized by applying optical high spatial resolution diagnostic techniques, as discussed below.

One of the useful tools for studying a discrete structure of the optical breakdown plasma column in liquids besides the Schlieren and shadowgraph approaches,[7] is Mach–Zehnder interferometry.[8] The technique has a high spatial (~20 μm) and temporal (≈6 ns) resolution. The region of the laser (1,064 nm) breakdown is tested by the second harmonic of a Nd:YAG laser system (532 nm). The time delay between the pulses of the lasers is controlled. Interferometric diagnostics using a Mach–Zehnder interferometer allows studying not only the structure, but also the dynamics of a laser plasma column in tap water and alcohols and investigating the mechanisms of optical breakdown.

The structure of the breakdown in tap water can be seen from a series of interferograms presented in Figure 1.4. A laser spark column consists of luminous plasma balls, bubbles arising over the plasma balls, and associated microspherical shock waves. Expanding vapor from the surface of an inclusion particle and vapors of water organize a bubble over the inclusion, which in its turn generates the powerful microspherical shock wave. The dynamics of the spherical shock waves are seen in Figures 1.4 and 1.5. In Figure 1.5, a laser spark column in ethanol is presented. Averaged Mach numbers of microspherical shock waves in tap water as well as in ethanol are presented in Figure 1.6 as a function of controllable time delay of the probing pulse.

All analyzed experimental data show that the mechanism of optical breakdown is based on inclusion particles, so that a laser spark column in tap water or ethanol is filled with microexplosions on separate inclusions. It is interesting to remark that in ethanol (see Figure 1.5[a]) a self-focusing channel is observed.

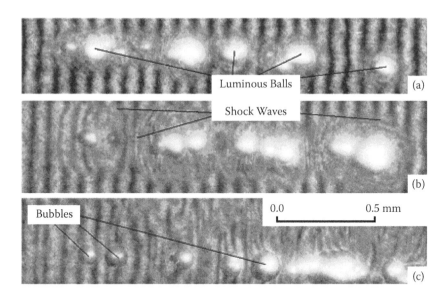

FIGURE 1.4
Interferograms of a laser spark column in tap water: (a) 17 ns, (b) 52 ns, and (c) 250 ns of controllable time delay.

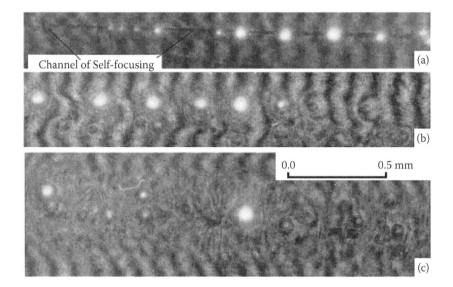

FIGURE 1.5
Interferograms of a laser spark column in ethanol: (a) 35 ns, (b) 64 ns, and (c) 226 ns of controllable time delay.

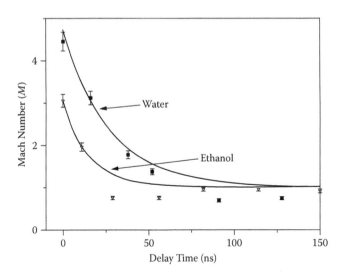

FIGURE 1.6
Dynamics of associated shock waves in tap water and ethanol.

2

Mach–Zehnder Reference Beam Optical Holographic Interferometry

2.1 Using Mach–Zehnder Interferometers for Recording Focused Image Holograms on Photographic Films

Note that a Mach–Zehnder interferometer (see Figure 1.1) can be successfully used as the basic setup for designing an optical system recording holograms with low spatial carrier frequencies, so called *Mach–Zehnder holograms*. Holograms recorded on a Mach–Zehnder interferometer require the implementation of spatial filtering during the process of reconstruction. Usually Mach–Zehnder focused image holograms recorded on photographic films have relatively low carrier spatial frequencies. The only difference in comparison with the recording interferograms is that the spatial frequency of Mach–Zehnder holograms is higher: $10 < f < 50$ lines × mm^{-1}. The background interference pattern on a hologram is called the *holographic fringe pattern* or *holographic grating*; fringes of the pattern are called *carrier* or *holographic*. The correspondent width of Mach–Zehnder holographic fringes is significantly smaller ($20 < d < 100$ microns) than the characteristic widths of Mach–Zehnder interferograms. Thus there is no qualitative difference between Mach–Zehnder holograms and finite-width-fringe interferograms, only quantitative. This quantitative difference, expressed in a spatial frequency of holographic fringes, was studied for the first time in References [12, 13]. This value is responsible for the procedure of optical filtering during reconstruction of Mach–Zehnder holograms, because their spatial carrier frequencies are not high enough to be reconstructed directly as in the case of ordinary optical holograms.

The interference fringe pattern of plane comparison and signal waves can be written in the form of Equations (1.4) and (1.5), where the spatial carrier frequency of a hologram, $f = \sin(\alpha)/\lambda$, depends on the offset angle α, and the laser line wavelength λ is exactly as it is in the case of interferograms. For example, if a hologram has the spatial frequency of holographic fringes ~25 mm^{-1}, and it was recorded by using of a helium-neon laser ($\lambda = 632.8$ nm), then the offset angle is determined as: $\alpha = \arcsin(f \times \lambda) \approx 0.9$ deg , which is

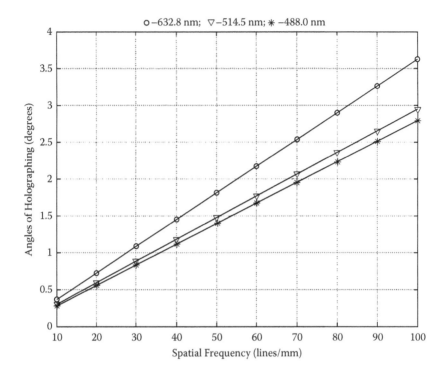

FIGURE 2.1
Calculated offset angles of Mach–Zehnder holograms as a function of spatial-carrier frequency.

remarkably larger than in the case of recording interferograms. Technically an enlargement of the spatial frequencies can be realized by rotating the beam-splitting cube (11) or the mirror (12) in Figure 1.1 over the vertical axis. The corresponding offset angles for the red line of a helium-neon laser and two lines of an argon ion laser are displayed in Figure 2.1.

It should be noted that although carrier spatial frequencies of Mach–Zehnder holograms are already higher than spatial frequencies of Mach–Zehnder interferograms, they are not high enough for reconstruction of the signal and comparison waves without overlapping neighboring orders, so that the separation of the wave(s) (collimated beams) without optical filtering will be impossible. Thus, methods of spatial filtering are necessary during the procedure of reconstruction of Mach–Zehnder holograms for the successful extracting needed diffraction order.

In Figure 2.1, where the dependence of the holographing angle as a function of the spatial frequency of Mach–Zehnder holograms is presented, it is seen that for holograms of 20 to 50 mm^{-1}, the angle of holographing is ~1 to 2 degrees.

The author has experimented with diverse types of commercial and technical photographic films to record high-quality, low spatial carrier frequency

Mach–Zehnder holograms. This experience has demonstrated that common commercial photographic films are able to produce holographic gratings of satisfactory quality with spatial frequencies ~20 to 50 lines × mm^{-1}, and has shown the usefulness of these photographic films for different Mach–Zehnder holographic interference methods.

Higher resolving powers of 35-mm photographic emulsions could guarantee recording Mach–Zehnder holograms having carrier frequencies in the range 50 to 75 lines × mm^{-1}. Kodak T-MAX 100, from Table 1.1, and orthochromic photographic films Kodak Rollei Ortho 25 (250 mm^{-1}) and Agfa Copex HDP (300 mm^{-1}) may be noted as examples. The photographic films mentioned are more suitable for reliably imaging fine details of phase objects to the point of the spatial resolution of a few microns. Such photographic films and others having the resolving power of 200 to 300 lines × mm^{-1} are also useful for the procedures of rerecording or rewriting holograms and, consequently, for performing interference measurements with enhanced sensitivity (see Section 3.2).

It is worth emphasizing that in this situation an imaging objective lens (lenses) should guarantee the high spatial resolution necessary for reconstruction of fine details of the phase objects under study. Experimental practice shows that common 35-mm photographic films and standard camera lenses (resolution of ~60 to 80 lines × mm^{-1}) may be useful tools in different Mach–Zehnder holographic applications in a wide range of magnifications.

The first experiments with low spatial carrier frequencies of 33.5 and 70 lines × mm^{-1} holograms were successfully performed[12,13] in the coherent light of low-power He-Ne gas lasers with offset angles of 1.2 and 2.5 deg. The holograms were recorded on Ilford FP4 plates and Tri-X Pan film, a relatively high-speed photo material with the exposure index of 400 ASA. Thus, it was shown experimentally that Mach–Zehnder holograms could be recorded on faster than standard photo materials.

A working range of the lowest carrier frequencies is governed by the following factors: the lowest border of the order of ≤10 lines × mm^{-1} is defined by the possibility of the reliable spatial filtering during the procedure of reconstruction and/or, in the case less stringent requirements, to the spatial resolution of fine details of the phase object in the image plane (the spatial resolution is ≤0.05 mm); the highest border of the order of ≤50 lines × mm^{-1} is defined by the limited frequency-contrast (modulation) characteristics of standard photographic films (~100 mm^{-1} resolving power) and the vignetting of the reference beam on an aperture of the imaging objective lens. It should also be noted that the procedure of obtaining quality film negatives with spatial frequencies ≥50 lines × mm^{-1} (using standard photo material) is cumbersome and requires experience with photography.

As already noted, Mach–Zehnder holograms with such relatively low spatial frequencies require the procedure of spatial filtering during the process of reconstruction. Usually the setup for reconstruction consists of the reconstructing collimated beam illuminating a hologram, an objective lens behind

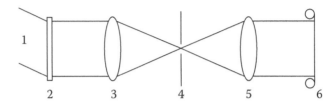

FIGURE 2.2
Reconstruction scheme for Mach–Zehnder holograms. (1) Reconstructing beam; (2) Mach–Zehnder hologram; (3–5) collimator, (4) stop diaphragm, (6) photographic film.

the hologram, and a one-hole diaphragm for selecting reconstructed order(s) of diffraction (see Figure 2.2). The hologram (2) as a holographic grating generates a system of collimated beams just behind its plane, which represent diffraction orders with the angular distance between them of $\alpha = \text{arcsine}(f \times \lambda)$ radians (in Figure 2.2, only a reconstructed signal (or comparison) wave is shown). The lens (3) with focal length F is used for selecting the diffraction orders at its back focal plane. The linear distance, s, between any two nearest circles of confusion in the focal plane, is expressed as follows [$\tan(\alpha) \approx \sin(\alpha) \approx \alpha$]:

$$s = F \times \alpha = F \times f \times \lambda \qquad (2.1)$$

For instance, if $f = 10$ mm^{-1}; $F = 200$ mm, and $\lambda = 632.8$ nm, then the calculated distance s is ≈ 1.26 mm. This is a pretty small distance, and the procedure of an object wave extraction should be performed very accurately. For lower spatial frequencies the procedure becomes unreliable from the point of view of overlapping the two circles of confusion. As was already stated, the value $f\sim 10$ mm^{-1} in fact defines the lowest border of the spatial frequencies of a Mach–Zehnder hologram.

2.1.1 35-mm Photographic Films for Mach–Zehnder Holographic Interferometry: Spatial-Carrier Frequency of a Hologram and the Finest Details of the Reconstructed Phase Object

From the theory[9] of reconstruction of plane amplitude thin holograms as thin amplitude gratings, it is known that the highest spatial frequency of the phase object variations (or the minimal resolvable detail) depends on the maximal spatial carrier frequency of the hologram. A hologram is thin if the thickness of emulsion is smaller than the spacing of the holographic fringes. For thin amplitude gratings the coefficient of amplitude transmittance changes with coordinate x as: $\tau(x) = <\tau> + \Delta\tau \cos(2\pi f)$, where $<\tau>$ is the average amplitude transmittance, and $\Delta\tau$ is the amplitude of the spatial variations.

A negative with the spatial frequency ≤ 50 lines $\times$ mm^{-1} can be treated as the plane grating. Indeed, the spatial frequency of ≤ 50 lines $\times$ mm^{-1} means that the grating's spacing is of the order of 20 microns or larger, and it is larger than the few microns of thickness of the photographic emulsion. If the value of spacing is getting closer to the thickness of an emulsion, the hologram begins to demonstrate characteristics of a volume holographic grating. A plane hologram is called "amplitude" or "phase" because of its determinative modulating parameter that affects the amplitude τ (x, y) [or intensity $T(x, y)$] transmittance coefficient of the hologram.

Analysis of the reconstruction of plane amplitude holograms[9] shows that the minimal carrier frequency f_{car} must be at least three times larger than the maximum spatial frequency, ξ_o, of the variations produced by the phase object under test: $f_{car} > 3 \xi_o$. Only in this case can zero and first orders be successfully spatially separated. As to the spatial frequency of the signal hologram in the area of the phase object, it may achieve the value $f_{sig} = f_{car} + \xi_o = 4 \xi_o$. For instance, if the minimal detail of a phase object under test is 0.02 mm (it should correspond to the spatial frequency $\xi_o = 25$ lines×mm^{-1}), then the carrier frequency should exceed 75 mm^{-1}, and the spatial frequency of the signal hologram should be ~100 lines×mm^{-1}. The frequency 100 lines×mm^{-1} must be resolved by the emulsion of a negative, and consequently, the resolution limit of the proper emulsion must be of the order of 100 mm^{-1}.

Thus, in regard to the resolving power, requirements for holographic photographic films are essentially harder (higher) than for photographic films, which are used for recording interferograms. It does not mean that the photographic films, indicated in Table 1.1 can be used only for producing interferograms. They are also applicable for producing holograms with spatial frequencies of, for example, 20 to 50 lines×mm^{-1}. From this point of view, to resolve the finest details of the object, the usage of photo materials with as high as possible resolving power in Mach–Zehnder holographic interferometry is more preferable. A good choice in this case are so-called photo films for microfilming. These types of contrast photo films, which meet high spatial resolution requirements of 100 to 300 mm^{-1} are ideally fit for Mach–Zehnder holographic interferometry even in the case of relatively high spatial frequencies. Examples of these films includes Agfapan 25, Isopan Fine Grain (Agfa), Technical Pan, and High Contrast (Kodak).

2.1.2 Resolving Power of the Ensemble of a Photographic Film Using an Imaging Lens

In classic interferometry, minimal resolvable detail of a phase object on the negative is determined mainly by the optical resolution of an imaging lens (an imaging collimator) of the Mach–Zehnder interferometer, provided that the resolution of the photographic film is higher. In the case of recording the Mach–Zehnder hologram, the resolving power of photographic film plays an appreciable role in separation of the first and zero orders and in recording a high spatial frequency of fine details of the phase object.

The case is that the system (lens–film) resolution R_S is a function of separate resolutions of the lens and the film:

$$\left(\frac{1}{R_S}\right)^2 = \left(\frac{1}{R_L}\right)^2 + \left(\frac{1}{R_F}\right)^2 \tag{2.2}$$

Here R_L and R_F are the resolving powers of the lens and the film, respectively. It is well known that resolution of most standard camera lenses does not exceed the value of 60 to 80 mm^{-1}. If the resolution of a lens is 50 and the film is 100 mm^{-1}, then the system only has the resolution $R_S \approx 45$ mm^{-1}. If the resolution of the lens $R_L = R_F = 100$ mm^{-1}, then the system resolution is $R_S \approx 71$ mm^{-1}. In any case, for classic or holographic interferometry the experimenter should choose the best achievable photographic camera lenses, because for singlets the resolving power is pretty low.

In addition, one should remember the following. If a high quality objective lens images the phase object through viewing windows having a relatively low optical quality, it is hardly possible to resolve fine details in the case of classic interferometry because of the fact that aberrations in classic interferometry are not cancelled. That is why it is highly recommended for the viewing windows of an evacuated chamber to use optics of the best possible quality. From the same point of view, the use of a quality objective lens instead of a conventional singlet is also absolutely necessary due to large optical aberrations of the latter.

2.1.3 Nonlinear Recording and Focused Image Holograms

Broadly speaking, a hologram has nonlinear characteristics of its coefficient of amplitude transmittance. As was noted in Section 1.2, the coefficient of amplitude transmittance for processed photo material should be written in the form of:

$$\tau(x,y) \approx I^{\frac{\gamma}{2}} \sim \left\{1 + \frac{a_o a_R}{a_o^2 + a_R^2}\left[\exp\left(i\Delta\Psi\right) + \exp\left(-i\Delta\Psi\right)\right]\right\}^{\left(-\frac{\gamma}{2}\right)} \tag{2.3}$$

Narrowly, nonlinear characteristics of photo materials are described in References [14, 15]. Equation 2.3 can be expressed as a Taylor series in the phase $\Delta\Psi$:

$$\tau(x,y) \approx 1 - \frac{\gamma}{2}\frac{a_o a_R}{a_o^2 + a_R^2}\left[\exp\left(i\Delta\Psi\right) + \exp\left(-i\Delta\Psi\right)\right] +$$

$$\frac{\gamma}{4}\left(\frac{\gamma}{2}+1\right)\left(\frac{a_o a_R}{a_o^2 + a_R^2}\right)^2\left[\exp\left(i\Delta\Psi\right) + \exp\left(-i\Delta\Psi\right)\right]^2 + \ldots$$

(2.4)

In optical holography, a hologram is recorded using a "linear" regime, which is characterized by a small object-to-reference beam ratio, $a_o/a_R \ll 1$. The linear recording helps to avoid the reconstruction of unwanted waves, which, when interfering with the object wave A_o, are able to deform it and cause the image fault. It can be seen from Equation (2.4) that in the case of a linear recording, intensities of waves in the second and higher orders are remarkably lower than in the first orders: $I_2/I_1 = [(\gamma+2)/4]\,(a_o/a_R)\,/\,[1+(a_o/a_R)^2] \gg a_o/a_R$. Thus the linearly recorded hologram is characterized by a low contrast. One may say that linear recording of holograms goes along with a low contrast of the holographic fringes and a low efficiency of diffraction in the first order. *Diffraction efficiency* (e_d) is defined as the ratio of the light intensity in the first diffraction order to the intensity of the reconstructing wave. Low diffraction efficiency in the first order can prevent studying the reconstructed waves by different optical techniques and accurately reducing interference data.

Unlike optical holography, in holographic interferometry there is no need for linear recording holograms. If the amplitudes of reference and object waves are the same, then the interference pattern has the maximum possible contrast and diffraction efficiency. The best results can be obtained if, first, signal and reference beams have the same intensity, and second, the phase object under study is sharply imaged in the plane of photographic film.

2.1.4 Focused Image Holograms

Here it would be pertinent to note some remarkable features of image holograms. Properly speaking, an object wave carrying useful phase information on the phase object under test also carries useless phase information on the region under test, for example, abrupt changes of the refractive index $n(x, y, z)$, which are also recorded on the signal hologram. Similar regions arise in the vicinity of a shock front, which is generated, for example, by a laser spark or a shock flow field over the model in a wind tunnel. These regions, characterized with high gradients of the refractive index, may more effectively refract diagnostic rays than the rest of a phase object. This effect leads to redistribution of a light intensity $I(x, y)$ in the image plane (or, in other words, in the plane of a hologram) in the form:

$$\frac{\Delta I(x,y)}{I(x,y)} \approx L \int_{z1}^{z2} \left(\frac{\partial^2}{\partial x^2} + \frac{\partial^2}{\partial y^2} \right) n(x,y,z)\,dz \qquad (2.5)$$

where L is the distance between the object and the hologram. In fact, the redistribution in Equation (2.5) is reminiscent of a shadow effect, at which the regions with abrupt changes of the second derivative of the refractive index are visualized[17,18] much better. The situation could be modeled by using the two object waves, which are recorded on a negative: the object nonrefracted wave A_o and the refracted wave $B = b \times \exp(i2\pi f_b x)$, where b is the amplitude of the refracted wave; $f_b = (\Delta a)/\lambda$ is the spatial frequency of the refracted wave, where Δa is the refractive angle. These two object waves interfere with a reference wave and with each other in the plane of a signal hologram. In accordance with Equation (2.4), after reconstruction two waves will propagate in the direction of the positive first order ($a_o = a_R = a$):

$$B \approx \frac{\gamma}{2} \frac{b/a}{1+(b/a)^2} \exp\left[i(\varepsilon - 2\pi f_b x) \right]; A_o \sim \left(\frac{\gamma}{2} \right) \frac{1}{2} \exp(i\varepsilon) \qquad (2.6)$$

It should be emphasized that the amplitude of the refracted wave, b, is comparable with the amplitude a only in the regions of a high second derivative: $b \approx a$; and it is assumed that $\Delta a \ll a$ or ($f_b \ll f$), where a is the holographing angle.

In the case of unfocused imaging of the object, a complicated interference picture caused by the two reconstructed waves could result in remarkable distortion of the reconstructed signal wave, mainly in regions of abrupt changes of the refractive index—so-called "shadow effects." The optical conjugation of an object and a signal hologram allows a notable simplification of the complicated interference picture, narrowing it down to a simple interference of the object and reference waves. In other words, focused image holograms make it possible to reconstruct the object wave without annoying distortions to the wave front, which arise during the reconstruction of the hologram, causing an unfocused image to be recorded. Interesting and applicable information on some features of image holograms can be found in References [16, 17].

2.2 Reconstruction of Focused Image Mach–Zehnder Holograms and Studying Signal Waves by Related Techniques

In contrast to photography, in holography, the amplitude transmittance coefficient τ plays a more important role than optical density D of the photographically processed photographic film. It is connected to the process of effective reconstruction of recorded wave(s). From this point of view it is more reasonable to use the densitometry relation of the amplitude transmittance coefficient versus exposures, i.e., (τ-H) instead of (D-H) dependence. Graphically imaged, this dependence is called the transmittance–exposure curve and is also presented in Reference [8]. This idea is related to the fact that the complex amplitude of any reconstructed object wave A_o is proportional to the amplitude transmittance[9] of a negative:

$$A_o(x, y) \approx \tau(x, y)A_{rec} \tag{2.7}$$

where A_{rec} is the reconstructing wave with the amplitude a_{rec}.

If one takes into account only the terms that are not higher than the first order, neglecting higher terms, the coefficient of amplitude transmittance at the condition, $a_o = a_R$, can be rewritten in the form:

$$t(x, y) \approx 1 - \frac{\gamma}{4}\left[\exp\left(i\Delta\Psi\right) + \exp\left(-i\Delta\Psi\right)\right] \tag{2.8}$$

Equation (2.8) clearly shows that when the coefficient of amplitude transmittance changes "harmonically" in the direction x, the hologram is an amplitude sinusoidal grating. If the hologram having the coefficient of amplitude transmittance in the form of Equation (2.8) is illuminated by a plane wave $A_{rec} \sim exp(i\Psi_R)$, then the optical field A_{of} behind the hologram consists of three waves only:

$$A_{of} \approx \exp\left(i\Psi_R\right) - \frac{\gamma}{4}\exp\left(i\Psi_o\right) - \frac{\gamma}{4}\exp\left[-i\left(\Psi_o - 2\Psi_R\right)\right] \tag{2.9}$$

This is not a surprising result, due to the sinusoidal behavior of the grating. Behind the hologram, only three diffraction orders are reconstructed: zero, positive first order, and negative first order. The simplified Equation (2.9) shows that the degree of amplitude modulation of the coefficient of amplitude transmittance and the intensity of the wave reconstructed in the positive (or negative) first order depends on frequency–contrast characteristics of the negative. The maximum possible modulation of the transmittance coefficient in the form of Equation (2.9) is equal to 1, if $\gamma = 2$. It can be seen from the equation that the amplitude of the first-order wave constitutes half of the amplitude of the zero-order wave. The ratio of the amplitudes of the reconstructed waves is $\frac{1}{2}$:$\frac{1}{4}$:$\frac{1}{4}$; the ratio of intensities is $\frac{1}{4}$:$\frac{1}{16}$:$\frac{1}{16}$, respectively.

The diffraction efficiency of a hologram was already defined as the ratio of an intensity of the wave, diffracted in the positive first order, to an intensity of the reconstructing wave. Thus the maximum diffraction efficiency of plane amplitude holograms is $e_d = 1/16 = 6.25\%$. Therefore, 25% of the light propagates in zero order, and 62.5% of the intensity of a reconstructing wave is absorbed in a sinusoidal amplitude grating.

The nonlinear-recorded hologram demonstrates a high contrast and a relatively high diffraction efficiency. A high intensity replica of the object wave reconstructed in positive first order is used for studying its front characteristics by different holographic techniques. Relatively high intensities of reconstructed waves in higher orders compared with linearly recorded holograms make it possible to use them for enhancing the sensitivity of interference measurements that will be shown in detail in Section 3.2.

Applications of Mach–Zehnder interferometers for recording holograms are inseparably connected to methods of spatial filtering at the stage of reconstruction. These methods allow the spatial selection of the needed diffraction orders in the back focal plane of a focusing lens; as to unwanted waves, a diaphragm (8) blocks them out (see Figure 2.3).

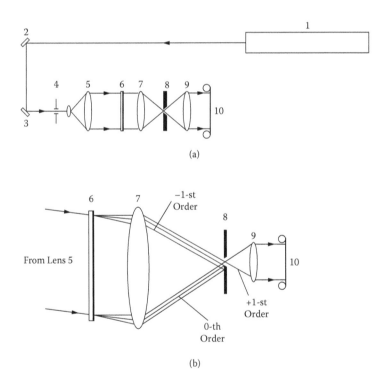

FIGURE 2.3
Optical scheme for reconstructing Mach–Zehnder holograms (a) and a close-up of spatial filtering. (b) (1) CW helium-neon or argon ion laser beam; (2, 3) flat R=100% mirrors; (4) spatial filter; (5, 9) collimating lenses; (6) hologram; (7, 9) collimator; (8) stop diaphragm; (10) photographic film.

A standard optical scheme for reconstructing a signal hologram and studying the signal wave is represented in Figure 2.3. A hologram (6), carrying phase information on the recorded object wave is illuminated by a parallel wave with unity amplitude, which models the reference wave. Its complex amplitude can be written as $A_R \approx \exp(\Psi_R)$. It is assumed that the reconstructing beam is free of optical aberrations because an ideal spherical wave behind the spatial filter (4) is collimated by a high-quality objective lens (5) (see Figure 2.3[a]).

Let us analyze Equation (2.9) in detail. The term $\exp(\Psi_R)$ is the zero-order wave traveling in the direction of the reconstructing wave. The second term, $\gamma/4 \times \exp(\Psi_o)$, is the positive first order wave, representing a reconstructed signal wave; and the third term, $-\gamma/4 \times \exp[-i(\Psi_o - 2\Psi_R)]$, is the negative first diffraction order. A stop diaphragm (8) located in the back focal plane of an objective lens (7) selects only the reconstructed signal wave, which travels in the normal direction. The waves corresponding to zero, negative first, and all higher orders are blocked out by the diaphragm (8).

Generally speaking, the reconstructed object wave carries not only useful phase information on the phase object under investigation, but also some phase distortions inherent to a scheme of recording φ_u, i.e., uncompensated phase errors due to imperfect optical elements, viewing windows, etc. (see Section 1.1.3). The reconstructed wave in the positive first-order signal wave with proper distortions and at the condition $\gamma = 2$ can be presented in the form:

$$\sim -\tfrac{1}{2} \exp[\varepsilon(x, y) + \varphi_u + \varphi_h] = -\tfrac{1}{2} \exp[\varepsilon(x, y) + \varphi_u + \varphi_h] \qquad (2.10)$$

In reality the reconstructed wave carries not only phase errors due to aberrations in the scheme of recording, but also aberrations in the hologram itself, φ_h. The source of these aberrations is emulsion shrinkage and a relatively low quality of the substrate such as acetate film or a glass plate. If all types of aberrations are negligibly small, then the second term in Equation (2.9) represents the replica of an original signal wave without any aberrations propagating normally to the hologram.

Finally, the reconstructed and selected replica of the signal wave can be studied by related coherent optical diagnostic methods. These techniques supplement reference beam holographic interferometry, which is analyzed below in Section 2.3. Optical diagnostic methods include shadow, Schlieren, Moiré, and reference beam dual hologram shearing interferometry techniques.

2.2.1 Holographic Shadow and Schlieren Techniques

Schlieren information represents the redistributed light intensity on a photographic film (a Schlierengram) due to the first derivative of the refractive index in the direction perpendicular to an optical knife-edge and can be easily extracted from a signal hologram.[18–19] This technique is frequently called holographic Schlieren. A deviation angle δ of light rays in the optical nonuniformity can be expressed in this case as:

$$\delta \approx \frac{1}{n(x,y,z)} \int_{z_1}^{z_2} \nabla_\perp n(x,y,z)dz \tag{2.11}$$

To record a Schlieren image, the stop diaphragm (8) in Figure 2.3(b) must be replaced with an adjustable optical slit. The width of the slit should be smaller than the distance between the nearest diffraction orders (±1st, zero) to perform an effective selection of the positive first order. An edge of the slit can be treated as the "optical knife-edge," which visualizes the first derivative of a phase object's refractive index in the direction perpendicular to the slit. To generate the Schlieren image, the edge of the slit should be placed in the back focal plane of the lens (7) and should partially overlap the circle of confusion that arises due to the focusing of the reconstructed replica. It should be remarked that, in the case of Schlieren measurements with *dark field*, the edge should fully overlap the circle of confusion in the positive first order. An exact location of the optical slit and its quality are critical to the characteristics of the Schlieren image focused on the photographic film (10). In Figure 2.3 the focusing objective lens (7) plays the role of the "Schlieren" lens.

Shadowgraphs provide information on second derivatives of the index of refraction. Unlike the Schlieren technique, the object wave is reconstructed without any stop diaphragm in the back focal plane of the focusing lens (7). Note that the existing stop diaphragm (8) in this case is used only in order to select the positive first order (i.e., the undistorted circle of confusion in the first order). The position of an image plane should be carefully adjusted for the appearance of *shadow effects*, which will be clearly observed at some position of the photographic film (10). For photographing the shadowgraph (i.e., the deformed signal wave), the optical scheme shown in Figure 2.3(b) can also be successfully used.

Examples of holographic shadowgraphs are presented in Figure 2.4. The image shown in (c) ($f = 40$ mm^{-1}) has more contrast and is sharp in comparison with the shadowgraph presented in (b), where it is a bit blurry because of the lower frequency ($f = 20$ mm^{-1}) of the carrier fringes and, consequently, the lower spatial resolution of the reconstructed image. In Figure 2.4(c) the shock, which is located at 0.68 mm from the spherical surface, can be clearly observed. The technique is sensitive to strong shocks only.

2.2.2 Holographic Moiré Deflectometry

The optical scheme of a holographic Moiré deflectometer is displayed in Figure 2.5(a). The replica of a reconstructed signal wave (1) is collimated by

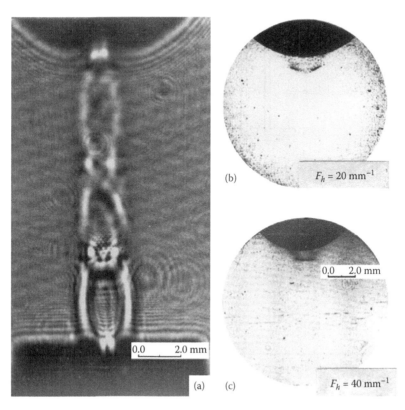

FIGURE 2.4
Classic (a) and holographic shadowgraphs (b, c) of supersonic air micro jet.[3,4] (a) Shadowgraph is recorded in the collimated beam of an argon ion laser (λ = 514.5 nm); (b, c) Shadow photographs reconstructed in scheme presented in Figure 2.3(b). Spatial carrier frequencies of holograms: (b) 20 and (c) 40 mm^{-1}.

the objective lens (2) and overfills the aperture of amplitude gratings (3, 4). The lens (2) optically conjugates an object and the plane of the second grating (4). The main principals of the design and operation of the Moiré deflectometer approach for wind tunnel testing is discussed in Chapter 8. The focusing objective lens (5) and the stop diaphragm (6) provide optical selection of one of the brightest diffraction orders from a complicated system of diffraction orders, which are generated behind the second amplitude grating. Without proper optical filtering (by the diaphragm), the reconstructed deflectogram would be of a low contrast and a low optical resolution due to the multibeam interference effect. Note that the collimator (5, 7) focuses the second Ronchi grating in the plane of the photographic film (8). One of the focused image deflectograms is presented in Figure 2.5(b). It is seen that the spatial resolution of the deflectogram is greater than the resolution of the correspondent reconstructed shearing interferograms (see next section).

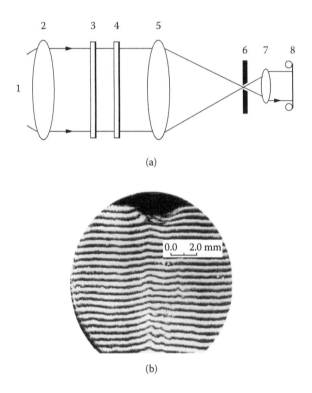

(a)

(b)

FIGURE 2.5
(a) Optical scheme of holographic Moiré deflectometry and (b) a reconstructed deflectogram of a supersonic air micro jet recorded on photographic film. (1) Selected signal wave; (2) collimating lens; (3, 4) Ronchi (amplitude) gratings; (5-7) collimator; (6) stop diaphragm; (8) photographic film.

2.2.3　Shearing Interferometry by Reconstruction of Two Reference Beam Signal Holograms

One of the useful and original optical diagnostic techniques applicable for testing phase objects recorded on Mach–Zehnder reference beam holograms is reference beam dual hologram lateral shearing interferometry. Usual shearing interferometry[20] is based on refraction of the diagnostic light due to gradients of the refractive index, which are perpendicular to the fringes. The interference picture arises as a result of the interference of two split diagnostic waves. If a shear between the sheared beams is small enough, the fringes in the lateral shearing interference picture visualize first differentials (gradients of the refraction index in the direction of the shear) of a phase object. Generally speaking, two shearing interferograms with mutually orthogonal fringe patterns are necessary for the complete evaluation of the wave front characteristics of a probing wave. Holographic interferometry makes it possible to simplify the complex optical scheme of classic shearing interferometers by means of replacing it with two optical elements (in mutually orthogonal directions)—identical reference beam signal holograms. Recording the two identical Mach–Zehnder reference beam holograms can be easy realized in the case of a stationary phase object. If the phase object is nonstationary, then the second holograms should be obtained by virtue of the process of duplication (optical copying).

The holograms are placed in a dual hologram holder. The proper optical scheme is presented in Figure 2.6(a). The same optical scheme will be applied below for demonstrating dual hologram technique. The dual hologram holder can produce not only two series of interference pictures with the mutually perpendicular shears, but also controllable lateral shear in any direction.

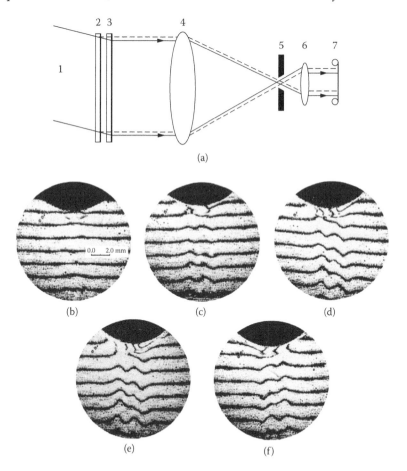

FIGURE 2.6
Optical scheme of reference beam dual hologram shearing interferometry (a) and reconstructed shearing interferograms (b–f) of a supersonic air micro jet in air recorded on photo film. (1) Reconstructing beam; (2, 3) reference beam signal holograms; (4, 6) collimator; (5) stop diaphragm; (7) photographic film. Spatial frequencies of holograms are ~20 mm^{-1}. Amplitudes of lateral shear: (b) 0.1; (c) 0.4; (d) 1.3; (e) 1.8; (f) –0.8 mm.

A reconstructing reference beam (1) with minimal aberrations illuminates two identical object reference beam holograms (2, 3). The holograms must be fixed in a dual hologram holder. The holder is a mechanical device, which permits one of the holograms to move independently with three degrees of precision control: rotation over the z-axis (optical axis) and translations in (x, y) horizontal and vertical directions. First of all, the signal holograms (2, 3) must be carefully adjusted to the point of complete coincidence; after that they are laterally sheared in some direction. The two identical signal waves are selected with a stop diaphragm (5), which is placed in the back focal plane of a lens (4); the diaphragm blocks all other waves. Only two object waves propagating in the positive first order are displaced in the figure; one of them is shown by the dashed line. A collimator (4, 6) images the plane of holograms on photographic film (7).

To obtain a shearing interferogram, one of the two holograms should be translated in the (x, y) plane with respect to the other hologram in order to generate the desired interference picture. The desired lateral shear spacing is regulated by means of translation in x or y directions. The orientation and spacing of a generated interferogram is dependant on spatial carrier frequencies and the orientation of holographic fringes. Some characteristics of shearing interferometry will be discussed in detail in Section II.

In Figure 2.6(b–f) shears were performed in the vertical direction. The five reconstructed shearing interferograms illustrate characteristics of the supersonic air micro jet with different amplitudes of shear and gradients of background fringe patterns.

In conclusion, it should be noted that the quality of an image produced by any of the presented visualizing techniques, excepting reference beam dual hologram shearing interferometry, strongly depends on the quality of the holograms or, in other words, on the quality of the reconstructed object wave. Optical distortions in the scheme of recording may disturb the picture in the image plane, which can lead to computational errors during the reduction of Schlieren, shadow, and Moiré interference data.

2.3   Studying Signal Waves by Methods of Reference Beam Mach–Zehnder Holographic Interferometry

The most powerful method of studying a reconstructed signal wave is interferometry. There are three main holographic interference methods: (1) double exposure; (2) dual holograms; and (3) real-time technique.[8] The choice is dictated by characteristics of the experimental setup and the phase object itself. Each technique has its advantages and limitations, but the most flexible and informative is the dual hologram method. Important methods of regulating the sensitivity of interference measurements are based on this technique.

2.3.1 Double Exposure Holographic Technique

The most simple and popular technique is the double exposure method. This approach is based on the exposure of the same photographic film two times: the first exposure is implemented without the phase object under test, and the second exposure, for example, occurs during the run of a facility with the object under study. The first holographic grating, recorded without the phase object, is called a *comparison* hologram; the second holographic grating, which is recorded on the same photographic film with a phase object, is called a *signal* hologram.

The optical scheme of reconstruction is shown in Figure 2.7. The design of the scheme is very close to the layout illustrated in Figure 2.2. A reconstructing collimated beam from the objective lens (5) generates a series of diffraction orders behind the double exposure hologram (6), which are spatially separated in the back focal plane of a focusing objective lens (7).

Two reconstructed positive first orders, i.e., signal and comparison waves (the comparison wave is shown with the dotted line), are reconstructed from the two holographic gratings (6) located in the plane of the same photographic film, selected by a diaphragm (8), and collimated by the objective lens (9). In Figure 2.7, only the signal and comparison waves selected by the diaphragm (8), are shown; all other waves are blocked out by the diaphragm. The focused reconstructed interference picture is imaged by means of the collimator (7, 9) in the plane of the photographic film (10). The telescope system (7, 9) optically conjugates the doubly exposed hologram (6) and the photographic film (10).

During the procedure of reconstruction, the comparison and signal waves are reconstructed with the same amplitude and propagate in the direction of the positive first order; namely, the comparison wave $\sim\exp[i(\Psi)_c]$ and the

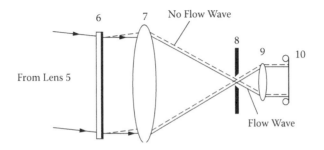

FIGURE 2.7

Optical scheme for reconstruction of double exposure holograms. (6) Double exposure hologram; (7, 9) collimator; (8) stop diaphragm; (10) photographic film.

signal wave $\sim\exp[(\Psi)_s]$. Taking into account all the discussed types of aberrations, the waves can be expressed as:

$$\exp[(\Psi)_c] = \exp\{i[\varphi_o + \varphi_h]\} \text{ and } \exp[(\Psi)_s] = \exp\{i[\varepsilon(x, y) + \varphi_o + \varphi_h]\} \quad (2.12)$$

The interference of the waves just behind the hologram is described as:

$$I(x, y) \approx 2 + \exp\{i[(\Psi)_s - (\Psi)_c]\} + \exp\{-i[(\Psi)_s - (\Psi)_c]\} = 2 + \exp[i\varepsilon(x, y)] + \\ \exp[-i\varepsilon(x, y)] = 1 + \cos[\varepsilon(x, y)] \quad (2.13)$$

We may expect that the contrast of the interference pattern in Equation (2.13) will be close to unity because both waves have the same intensity. The latter relation also shows that, in the interference pattern, recording and reconstructing distortions, φ_o and φ_h, are canceled.

Equation (2.13) represents an infinite-width-fringe interferogram. For obtaining a finite-width-fringe interference pattern it is necessary, in the middle of recording, to change the direction of a reference beam between the exposures. This leads to recording two holographic gratings with carrier fringes of different spatial frequencies (different holographing angles). During reconstruction it will lead to the appearance of a finite-width-fringe interference pattern.

In order to determine the width and direction of a background fringe pattern, and to demonstrate some important characteristics of the resulting interferogram, it is convenient to introduce the *vector of spatial frequency*, v. Its amplitude characterizes a magnitude of the carrier frequency of the holographic fringes, while the direction of the vector, which is perpendicular to the carrier fringes, shows the gradient of phase changes. This situation is illustrated in Figure 2.8.

Technically, it is most convenient to change a direction of the reference beam between exposures by rotating the beam-splitting cube (11) in Figure 1.1 over its vertical axis or the mirror (12). Note that changing a direction of the signal

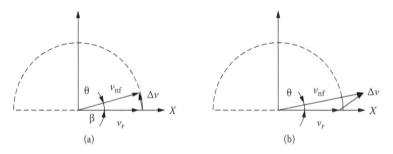

(a)   (b)

FIGURE 2.8
Graphical presentation of characteristics of resulting interference pattern during procedure of reconstruction of double exposure holograms. (a) Holographic grating having the same spatial frequency, $v_s = v_c$; (b) holographic fringes having different spatial frequency, $v_c \neq v_s$.

wave between exposures may lead to a "doubled" image, for example, a laser target etc. In Figure 2.8, the vectors of spatial frequencies for the signal and comparison holograms are determined as v_s and v_c, respectively. Generally speaking, if the correspondent holographic gratings are recorded with different spatial frequencies, the resulting vector, $\Delta v = v_c - v_s$, has the amplitude $\Delta v = [v_s^2 + v_c^2 - 2\,v_s\,v_c\cos(\beta)]^{1/2}$, where β is the angle between vectors. It is obvious that the direction and amplitude of the resulting vector Δv will definitely determine the spacing and position of the background fringes of the reconstructed interferogram. Let us consider two supreme cases: in the first case, amplitudes of the two vectors are of the same value (spatial frequencies of the gratings are the same), $v_s = v_c = v$, and β has an arbitrary value. The angle β is the product of rotating over the beam splitting cube (11) horizontal axis; in the second case, $\beta \approx 0$; $v_c \neq v_s$.

In the first case the resulting vector has amplitude $\Delta v = 2v\sin(\beta/2)$ and direction which is approximately perpendicular to both vectors. The finite-width-fringe interference background pattern has a horizontal system of fringes. For instance, if the vector v_c is rotated at 2 deg and its amplitude is 40 lines $\times$ mm^{-1}, then the interferogram has a spatial frequency of ~1.4 mm^{-1}. In the particular case, when $\beta = 0$, the resulting picture is an infinite-width-fringe interferogram.

Let us consider the second case, where the holographic gratings have different spatial frequencies and the magnitude of rotating the beam splitting-cube is of the same order, ~2 deg. The resulting amplitude of the spatial frequency vector Δv is $|v_s - v_c|$. The rotation of the beam-splitting cube at $\theta/2$ leads to changing of the offset angle, α: $\alpha + \theta$. The carrier spatial frequency will be $v_c = \sin(\alpha + \theta)/\lambda$, and the amplitude will be $\Delta v = \sin(\alpha + \theta)/\lambda - \sin(\alpha)/\lambda \approx v_s \sin(\theta)$. For $v_s = 40$ lines $\times$ mm^{-1} and $\theta = 2$ degrees, $\Delta v = 1.4$ mm^{-1}; that means $v_c = 41.4$ lines $\times$ mm^{-1} (or 38.6 mm^{-1}, if the spatial frequency of the no-flow holographic gratings is smaller). If both vectors are directed horizontally ($\beta = 0$), the system of interference background fringes is vertical. If the angle is not equal to zero ($\beta \neq 0$), then the system of background interference fringes is turned away from its vertical position. In conclusion, it should be noted that the double exposure technique provides the ability to arrange practically any desirable interference background fringe pattern when having only a series of holograms with a vertical system of carrier fringes.

2.3.2 Dual Hologram Technique

One of the limits of the double exposure technique is that with one double exposure hologram, one may reconstruct only one interferogram with a fixed system of background fringes. This limit can easily be overcome by reconstructing the interference picture from a pair of signal and comparison holograms recorded on separate photographic films.

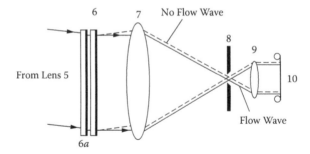

FIGURE 2.9
Dual hologram reconstruction scheme. (6, 6a) Signal (flow) and comparison (no-flow) holograms; (7, 9) collimator; (8) stop diaphragm; (10) photographic film.

The idea of the technique is based on the preliminary recording of signal and comparison waves on separate photo films, and subsequent interference comparison of them by using a dual hologram holder. In this instance proper variations of spacing and directions of the background interference fringes could be easily performed during the procedure of reconstruction simply by moving and rotating one of the holograms.

The principal distinction in comparison with the double exposure technique is that a series of a single infinite and many finite-width-fringe interferograms with arbitrary directions and spacing of background fringe patterns can be obtained during the procedure of reconstruction using only one signal hologram. Such an approach is very useful in cases of a unique phase object, unsatisfied reproducibility of a phase object from run to run, or for instance, in the case of studying the turbulent phase object when two reconstructed waves from the two signal holograms are similar only statistically. Examples of these objects will be given below.

A dual hologram reconstruction scheme is presented in Figure 2.9. A dual hologram holder is used for experimental realization of the qualities of holographic interferometry. As was noted, the holder is a mechanical device which permits both holograms to be independently moved with three degrees of precision control: translations in horizontal and vertical directions and rotation over the optical axis.

A collimated reconstructing beam from a lens illuminates the signal and comparison holograms (6, 6a). A series of diffraction orders reconstructed from the holograms is spatially separated in the back focal plane of an objective lens (7). Selected signal and comparison waves travel in the direction of the positive first order; the other orders are blocked out. Only these two positive first-order waves are collimated by an objective lens (9) and sharply focused on a photographic film (10). The photographic film can be replaced with the CCD/CMOS sensor for the purpose of acquiring the interference picture, which arises as a result of interference of the reconstructed signal and comparison waves, and transferring it to a hard disk (see Section III).

The interference picture is formed in accordance with Equations (2.12) and (2.13) in just the same way as in the double exposure technique.

Using only a single flow (signal) hologram, the technique is able to generate any necessary number of reconstructed interferograms, with arbitrary orientation of fringes and their widths, during the process of reconstruction by means of translation and rotation of different no-flow holograms in the dual hologram holder.

As was already discussed, a series of interferograms with the horizontal system of background interference fringe but with different spatial frequencies could be obtained having only a pair of flow and no-flow holograms with the same spatial frequency of the vertical system of holographic fringes. The reconstruction of a series of interferograms with vertical background fringes and different widths requires a series of comparison holograms with different spatial frequencies (it is assumed that the interferometer's setup is located in the horizontal plane for both cases). It is useful to note that if there is some sort of target, for example, an aerodynamic target in the field of view (or a laser target), then the comparison hologram should be exposed without it. Otherwise the image of the model would be doubled, so that, for example, information on boundary layers could be lost.

One of the remarkable characteristics of dual hologram (and double exposure technique) interferometry is that it is differential in time. This means that two consistent in-time phases of the object's variations can be compared interferometrically. Moreover, two different signal waves, which are reconstructed from two different signal holograms (or from two holographic gratings recorded on the same photo material), obtained under different conditions of the experiment, may also be compared interferometrically. As an example, a phase object could be laser spark initiated at two different gas pressures and/or in two different types of gases, or even by two different laser wavelengths from two different lasers. As a result of interference of these waves, an interference picture appears with the phase change difference $\Delta\varepsilon(x, y, z) = \varepsilon_1(x, y, z) - \varepsilon_2(x, y, z)$. The second interesting application is an interferometric comparison of two images of the same turbulent object in the same region of space at different periods of time.

One of the remarkable features of the dual hologram technique is that any signal hologram with an object wave recorded on it can be used for studying the phase object by Schlieren, shadow, Moiré, and dual reference beam hologram shearing techniques, as it has been stated and demonstrated above.

2.3.3 Real-Time Holographic Interference Technique

In the double exposure technique, signal and comparison waves are reconstructed from the two holograms recorded on the same photographic film. In the dual hologram technique the waves are reconstructed from two separate holograms. In this section the third situation will be discussed, in which a comparison wave is reconstructed from a comparison hologram by means of

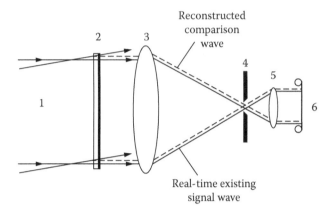

FIGURE 2.10
Real-time holographic interferometry scheme. (1) Reference wave; (2) comparison hologram; (3, 5) collimator; (4) stop diaphragm; (6) photographic film. The dashed line shows comparison wave.

the reference wave, whereas the signal wave propagates through the object under test and exists in real time during the run of a facility. These two waves generate "live fringe" interferometric pictures, since the signal wave permanently changes its phase during temporal evolution of the object under investigation. The optical scheme of the real-time interference technique is presented in Figure 2.10.

A focused image comparison hologram (2), with a comparison wave recorded on it, is photochemically processed after exposure and relocated at its original place. A reference wave (1) generates the replica of the comparison wave behind the comparison hologram (2) in the direction of the first positive order. It is assumed that the comparison wave propagates in the normal direction. A Mach–Zehnder or any other more complicated holographic interferometer could be used as a recording optical scheme. The only condition is that the interferometer should allow real-time phase objects to be studied. It is supposed that the size of the phase object does not exceed an aperture of the comparison hologram. The second wave participating in generation of an interference picture (6) arises when a real time signal wave, penetrating at first the phase object (not shown in the figure) and carrying useful phase information on the object under study, penetrates the comparison hologram in the normal direction. If the signal wave propagates perpendicular to the plane of the hologram, it travels together with the reconstructed comparison wave in the positive first order. A stop diaphragm (4) selects only the signal and comparison waves and blocks out other waves. Only these two waves are shown in the figure behind the hologram. The infinite-width-fringe interference picture of *"living fringe"* is imaged on the photographic film (6) by the collimator (3, 5). Technically, the phase object, located not far away from the comparison hologram, and the photographic film and the plate are lying in optically conjugated planes.

Taking into account the condition $a_o=a_R$, the coefficient of amplitude transmittance of the comparison hologram can be presented as follows:

$$\tau_c = 1 - \tfrac{1}{2}\{\exp[i(\varphi_o + \varphi_h - 2\pi fx)] + \exp[-i(\varphi_o + \varphi_h - 2\pi fx)]\} \qquad (2.14)$$

Reconstructed by the reference wave $\exp(-2\pi fx)$, the comparison wave is expressed as $\tfrac{1}{2}\exp(i\varphi_o)\exp(i\varphi_h)$. The signal wave just behind the comparison hologram can be written in the form: $\exp\{i[\varepsilon(x, y)+ \varphi_o+ \varphi_h]\}$ because, with respect to the comparison hologram, the real-time object wave diffracts in the plus first order. Thus, two waves propagate behind the comparison hologram in the same direction :

$$\tfrac{1}{2}\exp\{i[\varphi_o+\varphi_h]\} \text{ and } \exp\{i[\varepsilon(x, y)+ \varphi_o+ \varphi_h]\} \qquad (2.15)$$

The resulting infinite-width-fringe interference pattern can be written as:

$$I(x, y) = 1 + 4/5 \times \cos[(\varepsilon(x, y)] \qquad (2.16)$$

Equation (2.16) shows that the contrast of the resulting interference pattern has the maximum value of 80%. Consequently, in the case of the real-time interference technique, the contrast is a bit lower than in the case of the double exposure or dual hologram methods, where the contrast is close to unity. This is caused by the two interfering waves that have different amplitudes ($\tfrac{1}{2}$ and 1), whereas in the double exposure and dual hologram techniques the comparison and signal reconstructed waves have exactly the same amplitude. In the presented analysis it was assumed that the object and reference beams at the output of the Mach–Zehnder interferometer have the same intensity. The ideology and examples of applying the real-time shearing holographic interference technique are given in Section II.

Besides the limitation connected to the lower contrast of the interference picture, another limit concerns relatively low diffraction efficiency of plane amplitude holograms. It was shown above that their diffraction efficiency does not exceed 6.25%. It is well known[8,9] that thin-phase holograms have higher diffraction efficiency, easily exceeding 20% (the theoretical limit is 33.8%). Thus using a thin-phase hologram instead of a thin-amplitude one is more attractive for real-time interferometry, so far as a comparison hologram generates the first-order wave with a higher intensity.

Higher-diffraction-efficiency phase comparison holograms can be obtained after the procedure of bleaching, when the metallic silver grains in the photo emulsion are converted to transparent silver salts. The refraction index of these salts differs from the index of refraction of the gelatin matrix. For example, the indices of refraction of silver bromide and silver chloride[8] are 2.25 and 2.07, respectively. Amplitutde holograms with a high optical

TABLE 2.1

Bleaching Solution

Substance	Quantity
H_2O (distilled water)	600 milliliters
$CuSO_4$ (copper sulfate)	100 grams
NaCl (sodium chloride)	100 grams
10% H_2SO_4 (sulfuric acid)	250 milliliters
H_2O (distilled water)	until 1 liter

density would have a good quality—at least $D \geq 2$. This bleaching procedure should be performed with definite precautions,[8] because some bleaching solutions are hazardous for human health. For a long time, the author used the solution with components indicated in Table 2.1. This bleaching solution is safe, stable, and has a long lifetime. The bleached photo plates are transparent with a light blue undertone.

3

Dual Hologram Interferometry with Enhanced Sensitivity

3.1 Applications of Mach–Zehnder Holographic Interferometry

In this section, some results of Mach–Zehnder holographic interferometry operating with low carrier frequency holograms applied to four different phase objects under investigation are discussed. Focused image holograms were recorded on photographic films having resolving powers of 100 to 300 lines × mm^{-1}. The carrier spatial frequencies of the holograms were varied from 20 to 50 mm^{-1}.

3.1.1 Supersonic Free and Impinging Spherical Target Air Micro Jet

In the experiment described in Section 1.3 and illustrated in Figure 1.2, the pictures of an air micro jet[4] clarify the characteristic view of the phase object.[3] Note that the holographic recording also allows study of the micro jet by a few diagnostic techniques. Indeed, the holographic approach makes it possible to obtain principal information on the phase object under test due to subsequent reconstruction and the comprehensive study of a signal wave.

The dual hologram method is a versatile and comprehensive technique because it does not have some of the limitations of other interference methods. The technique was used to accurately measure the density distribution in the supersonic air micro jet as a function of its radius. The Mach–Zehnder scheme[3] was used for making Mach–Zehnder 20 to 50 mm^{-1} holograms on 200 lines × mm^{-1} resolving power 35-mm orthochromatic photocopying film. A continuous wave (CW) argon ion Beam-Lock 2060 laser emitting a green light at the wavelength 514.5 nm and at the intensity ~50 mW was used as the source of coherent radiation. The same laser line was used in the schemes of reconstruction. The maximal field of view was 25.2 × 16.8 mm^2, and the coefficient of magnification was ~1.4×. The spatial resolution of the optical scheme was about 0.04 mm. The typical reconstructed interferograms are shown in Figure 3.1 and Figure 3.2.

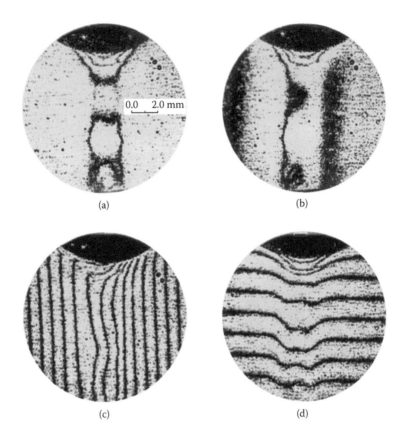

FIGURE 3.1
Reconstructed interferograms of supersonic air micro jet. (a) Infinite- and (b–d) finite-width-fringe interferograms; $f_{car} \approx 20$ mm^{-1}.

All eight interferograms were reconstructed from a pair of signal holograms with $f = 20$ and 40 mm^{-1}, having vertical systems of carrier fringes. The object and reference arms of the Mach–Zehnder interferometer are located in the horizontal plane. It already was noted that, having a series of holograms with vertical holographic fringes, it is possible to generate any number of reconstructed interferograms with arbitrary directions and spacing of background interference fringes.

The interferograms in Figure 3.1 were obtained by using only one signal and three comparison holograms of different spatial frequencies, all having a vertical system of holographic fringes. The interferograms in Figure 3.1(a) and (d) were reconstructed from a pair of comparison and signal holograms with carrier fringes of the same frequencies. Interferograms in Figure 3.1(b) and (c) were obtained by using two other comparison holograms, having holographic fringes with spatial frequencies that differ from the frequency of the signal hologram.

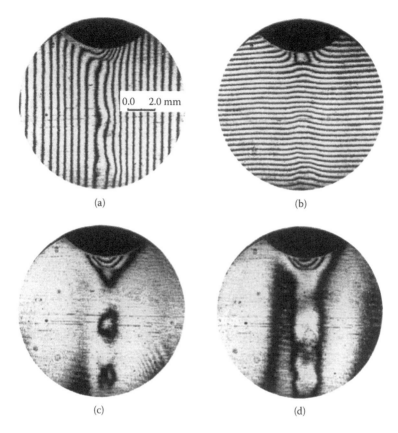

FIGURE 3.2
Reconstructed interferograms of supersonic air micro jet. (a, b, d) Finite- and (c) infinite-width-fringe interferograms; $f_{car} \approx 40$ mm^{-1}.

All reconstructed interferograms presented in Figure 3.2 are also obtained from one signal hologram. The interferograms in Figure 3.2(b) and (c) are obtained the same way as in Figure 3.1(a) and (d); a relatively small spacing of the background interference fringes in Figure 3.2(b) was obtained by rotating the comparison hologram over the z-axis at a remarkable angle.

The finite-width-fringe interferograms (Figure 3.1[b] and Figure 3.2[d]) are reconstructed by using comparison holograms with carrier frequencies, which slightly differ from the frequency of the signal holograms. It can be seen from the figures that the background system of interference fringes could have a positive or negative phase gradient depending on the relative position of the signal and comparison waves (in the scheme of the holographic interferometer).

The jet is 11 mm long before impinging the spherical surface. A periodic structure of "barrels" is observed in this region. All the interferograms show the periodic nature of barrels formation, which indicates a supersonic regime of free releasing the micro jet of air at the normal atmosphere pressure. The Mach number was determined as M ≈ 1.15.

The visual analysis of interferograms reconstructed from 20 to 50 mm^{-1} holographic gratings shows that the minimal resolvable nonuniformity is defined by the spatial carrier frequency of the gratings. It was experimentally found that reconstructed interferograms possess the spatial resolution which is close to the spatial resolution of the optical scheme of recording if the holographic carrier frequencies exceed ~40 mm^{-1}.

It is seen in the two series of interferograms that reconstruction from ~40 mm^{-1} holographic gratings leads to more contrast images with higher spatial resolutions.

One of the main problems in classic interferometry is the basic lack of continuous interference data.[21] In any given image, only a limited number of fringes cross any line along which quantitative data are desirable; therefore, they are obtained in a series of runs. The dual hologram technique can overcome this difficulty by manipulating the two holograms in a dual hologram holder. It is possible to maneuver the fringes in a controlled fashion to get a shift at any point desired in the field.

3.1.2 Vaporization of Suspended Volatile Liquid Droplets in Still Air

For the first time, the dual hologram technique was applied in Reference [22], wherein optical diagnostics were conducted on highly volatile liquid droplets with diameters of the order of 1 to 2 mm vaporizing in still air. The technique was used to precisely measure a change in the refractive index over droplets caused by droplet vaporization. This method is also able to provide a high spatial resolution (<0.05 mm), and is sensitive enough to measure the vapor density far away (a few droplet radii) from the surface of a droplet.

In order to increase the temperature of a droplet, and therefore the vapor density, a stainless steel fiber with a diameter of 0.55 mm and a high thermal conductivity was used for suspending droplets. As the source of coherent radiation, continuous wave helium–neon and argon ion lasers were used, with wavelengths of 632.8 and 514.5 nm, respectively. The dual hologram technique allowed reliable measuring of a fringe shift up to $^{1}/_{20}$, which is important in the case of thin phase objects such as vapors over volatile liquid droplets. Higher than the standard accuracy (0.1 fringes) might be achieved owing to compensations of optical aberration of the scheme of recording Mach–Zehnder holograms. In any case, the output distortions are smaller than $2\pi/20$. For the typical optical path difference of the order of ~λ, the accuracy equals 5%. The problems of weak phase objects and the accuracy of their interference measurements will be discussed in Section 3.2 in detail.

The unique feature of the technique is that a series of no-droplet (comparison) and droplet (signal) holograms are recorded at different times on separate holograms. The signal and comparison waves are reconstructed by using a dual hologram holder. By precisely varying orientation of the two

holograms, one may control the fringe configuration of a reconstructed interferogram. This capability permits the formation of a series of interferograms
with any background fringe spacing and orientation desired. The horizontal system of the background interference fringes, which are perpendicular to a suspending fiber, allows calculation of the radial distribution of the
refractive index of the air–vapor mixture for every cross section in the axial
direction. The schematic picture of diagnostics of a volatile liquid droplet is
presented in Figure 3.3(a).

The Mach–Zehnder scheme was used for making low spatial frequency
(up to ~50 mm^{-1}) holograms on Agfapan APX 25 (100 mm^{-1} resolving power)
and on Agfa microfilm (300 mm^{-1} resolving power). One of the reconstructed
interferograms with interference fringes perpendicular to the suspending
fiber is shown in Figure 3.3(b).

The dynamics of vaporization were investigated by holographing a vaporizing droplet with 6 s of minimal time between frames. Reconstructed holograms yield a series of interferograms with temporally varying parameters
of a droplet. Each interferogram was analyzed and used to determine a
radial distribution of changes of the refractive index of an air–vapor mixture
and a droplet size.

The radial distributions of $\Delta n = n(r) - n_\infty$ versus the radius r were calculated
for the equatorial cross sections. The values of Δn were found with 10% accuracy for acetone and with better accuracy for chloroform and diethyl ether
because of their larger index of refraction.

Radial distributions of $\Delta n(r)$ in the vicinity of a droplet surface, obtained
for all three liquids in the light of a helium–neon laser with a wavelength of

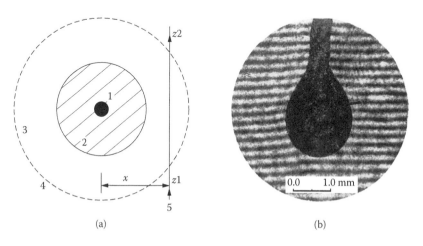

(a)   (b)

FIGURE 3.3
Mach–Zehnder holographic interferometry of a vaporizing volatile liquid droplet. (a) Top view
and (b) typical reconstructed interferogram of a droplet, suspended on a hollow metal fiber.
(1) Suspending fiber; (2) droplet; (3) vapor; (4) conventional boundary of vapor; (5) diagnostic
light ray.

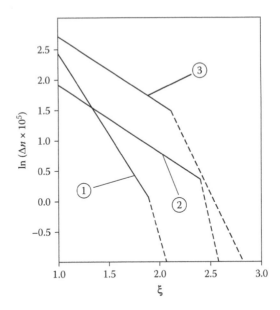

FIGURE 3.4
Radial distribution of the refractive index of vapor-air mixture over a droplet versus dimensionless variable: $\xi = r/r_d$; r is radius; r_d is radius of droplet. (1) Chloroform; (2) acetone; (3) diethyl ether.

$\lambda = 632.8$ nm, are shown in Figure 3.4. From numerous reconstructed interferograms, droplets having radii of $r_d = 930 \pm 50$ µm were chosen for all three substances. The values $\Delta n = n(r) - n_\infty$ were plotted semilogarithmically versus the dimensionless variable $\xi = r/r_d$, where r_d is the radius of a droplet.

The obtained interference information enables determination of the index of refraction, the vapor density, and the saturation pressure on the droplets' surface. Taking into account theoretical temperature profiles around droplets, one may calculate radial distributions of vapor density for all three substances.[22]

3.1.3 Thermal Waves over Channels in Plexiglas Drilled by CO_2 Laser

The experimental situation is described in Section 1.2. We have implemented a relatively soft regime of drilling Plexiglas.[7] A carbon dioxide pulse periodic laser with parameters of a laser pulse ($\varepsilon = 40$ mJ; $\tau = 20$ µs) allowed drilling at values of the intensity and energy density close to optimal ones: $q = 1.2$ MW $\times$ cm^{-2} and $E_s = 25$ J $\times$ cm^{-2}. As this occurs, the size of the area of irreversible change of the refractive index does not exceed 0.02 mm in the radial direction. In this case, the drilling capacity was increased to 85 µg/J^{-1}, and a conic channel converged more smoothly.

In one of the drilling regimes examined here, the process of space–time relaxation of a thermal field just after the affecting train of ten CO_2 laser pulses, with the time interval between the pulses of 2/3 s was investigated. The reconstructed interferograms for different depths (lengths) of the laser channels are presented in Figure 3.5.

The temperature and the change of the refractive index of the plastic are related by the expression: $\Delta n = (\partial n / \partial T)_p \Delta T$ at each point of the thermal field. The constant $(\partial n / \partial T)_p$ is equal to 9×10^{-5} K^{-1}. Accordingly, the calculated magnitude of the refraction index $\Delta n \approx 10^{-3}$, corresponding to the heating of Plexiglas near the wall of the channel (Figure 3.5[b]) is $\Delta T \approx 11$ K.

Visual analysis of the interferograms shows that, for the shortest channels less than 1.5 mm, the thermal wave is about spherical symmetry (Figure 3.5[a]), which is replaced by cylindrical symmetry for longer channels (Figure 3.5[b,c]). The infinite-width-fringe interferograms (Figure 3.5[a,c]) are reconstructed from the signal and comparison holograms, having carrier fringes of the same spatial frequencies. To obtain the finite-width-fringe interferogram (Figure 3.5[b]), the comparison hologram must have the smaller (bigger) frequency of carrier fringes. The vertical background interference fringes are obtained due to the vertical holographic fringe in the signal and comparison holograms.

In Figure 3.6, interferograms of the thermal waves for two different numbers of pulses in trains (5 and 25 pulses), just after finishing the trains, are presented.

The infinite-width-fringe interferograms are reconstructed by using the comparison and signal holograms having the same frequency of carrier fringes. The holographic fringes are vertical for both holograms.

It was easy to determine the quantity of heat released on laser drilling for long channels by using temperature fields computed from the interference patterns. Graphical integration allows us to evaluate the fraction of energy carried away by a heat wave for the laser pulses of ~100 mJ (see Figure 3.5[b]) taking into account the specific heat of Plexiglas, 1.46 J $\times$ g^{-1} $\times$ K^{-1}. Thus, the relative contribution of the thermal energy was approximately (100 mJ)/(40 mJ $\times$ 10) = 25%. For longer channels (Figure 3.5[c] and Figure 3.6) the coefficient of conversion is larger, and for ~6-mm channels exceeds 40%.

In conclusion, it would be pertinent to note that the thin (~2 mm) Plexiglas slab, which was cut from an ordinary plastic sheet, as an optical element of the interferometer has a very low optical quality and was scratched. Numerical evaluations on an interferometric picture (see Section 1.3) show that the nonflatness of the plastic slab may exceed ~20 λ, where λ is the wavelength of diagnostic light. Nevertheless, the dual hologram technique is able to perform fine optical measurements with the accuracy of ~ λ/10 due to canceling optical aberrations.

3.1.4 Visualization of Acoustic Waves in Dye Solution

This section reports on the first use of holographic interferometry to study the generation and dynamics of finite amplitude acoustic waves in alcohol.

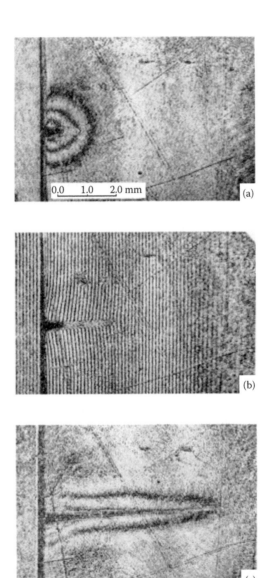

FIGURE 3.5
Reconstructed interferograms of thermal waves propagating in the area surrounding a laser channel created by a single pulse of a CO_2 laser; (a) and (c) infinite-width-fringe, and; (b) finite-width-fringe interferograms. Length of channels: (a) 1.2; (b) 2.6; and (c) 6.0 mm. Time delay: (a) 5; (b) 7; and (c) 15 sec; resolution <50 μm.

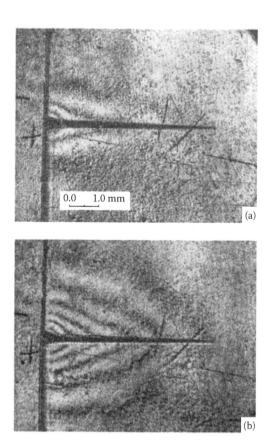

FIGURE 3.6
Reconstructed interferograms illustrating dynamics of thermal waves propagating away from a laser channel in the radial direction. (a) Five-pulse train; (b) 25-pulse train of CO_2 laser; length of channel is ~6 mm.

These accoustic waves appear when the beam from a second harmonic of a ruby laser is focused onto the cuvette boundary, which holds a dye solution: ethanol solution of crypto cyanine.[23,24]

The experimental layout includes the pulsed ruby laser with parameters: $\lambda_r = 0.694.3$ nm, $E_r \sim 1$ J, $\tau_r = 50$ ns. The second harmonics from the ruby laser are used to pump a dye laser with a diffraction grating positioned at the glancing angle of 87 deg in the resonator and having the parameter values: $\lambda_d = 580$ nm, $E_d = 0.5$ mJ, and the coherence length ~1 cm. This beam was used for probing acoustical waves in the radial direction. Part of the light from the first harmonic (~50 mJ) was intercepted by a beam-splitter and focused by a lens on the liquid–glass interface of the cell. The cell was placed in the object arm of a Mach–Zehnder interferometer.

The probing beam from the dye laser was directed into a multiple-mirror optical delay line, making it possible to adjust the time interval between the exiting and probing light pulses up to 200 ns. During the absorption of the exiting light in the dye solution, an axisymmetric heated region with characteristic diameter ~0.25 mm and length ~1 mm was formed in the region of the focal spot. The energy evolved in the liquid volume was ~1 kJ × cm^{-3}. The energy absorbed by the dye solution caused a sharp heating in the absorption zone up to a temperature ~250°C and an increase in the pressure in this region up to ~5 kbar.

The sharp increase in the pressure leads to the formation of a cylindrical acoustic wave, which propagates away from the heated zone, and the rarefaction wave, which propagates into the heated zone.

Holographic interferometry enables the change caused in the refractive index of the liquid by the increase in a temperature in the heated zone to be measured, as well as the change in the density of the medium in the region of the acoustic wave. The reconstructed focused image interferograms in Figure 3.7(a–c) show the time evolution of the zone of interaction. For the interferogram Figure 3.7(a) there is no time delay between the maxima of the probing and exiting pulses. In this case, the change in the refractive index is the exclusive result from heating the liquid: $\Delta n = (\partial n/\partial T)_p \Delta T$. The interferograms in Figure 3.7(b) and Figure 3.7(c,d) show the change in the refractive index for delays of 100 and 200 ns, respectively.

It can be seen from the interferograms that a cylindrical acoustic wave of finite amplitude arises at the boundary between heated and cold regions of the liquid and propagates away in the radial direction. The propagation velocity of this wave can be estimated from the interferograms as 1.5 km/s; the speed of sound in ethanol is ~1.16 km/s. Over the time of 200 ns, the acoustic wave has not yet acquired a sharp front; therefore the interference fringes in the zone of compression of the liquid have a smooth curvature as shown in Figure 3.7(d).

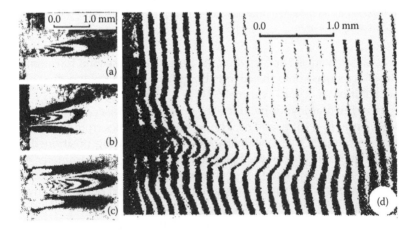

FIGURE 3.7
Reconstructed interferograms of heated zone and cylindrical acoustic wave in dye solution; (a–c) infinite-, (d) finite-width-fringe interferograms. Time delay: (a) 0, (b) 100, (c) and (d) 200 ns.

3.2 Dual Hologram Interferometry with Enhanced Sensitivity

There is a class of challenging phase objects that gives rise to very small fringe shifts in the resulting classic or reconstructed interference pattern. Some examples of such objects are: an ablation laser plasma or gas jet expanding into vacuum or a low-pressure gas atmosphere, plasmas of a low degree of ionization, and hypersonic shock flows of low densities over models in hypersonic shock tunnel facilities. They could cause very small fringe shifts, smaller than $\Delta N = 0.1$ (i.e., smaller than the visually unresolved fringe shift $\Delta N_v = 0.1$). In this section, it will be shown that such phase objects can be successfully tested by applying the Mach–Zehnder holographic approach to enhancing sensitivity of interference measurements.

3.2.1 Interference Measurements of Thin Phase Objects

The possibility of measuring thin (weak) phase objects by applying classic and holographic interferometry approaches requires special analysis. It is clear that, in the case of classic interferometry, the resulting fringe shift at the output of the Mach–Zehnder interferometer consists not only of the useful phase changes $\varepsilon(x, y)$, but also of the unwanted distortions $\varphi_u = \varphi_o - \varphi_R$. The level of unwanted fringe shifts determines the minimal measurable useful phase changes $\varepsilon(x, y)$. The real value of unwanted distortions φ_u can be found in the interferometric picture obtained without the phase object under study. The magnitude of such deviations should be smaller (preferably much smaller) than the visually unresolved fringe shift. If the fringe shifts due to phase objects do not exceed the level of unwanted distortion $2\pi/10$ (or are of the same order), then phase measurements of the phase object cannot be performed.

At the same time application of the Mach–Zehnder holographic technique guarantees measuring thin phase objects with any reasonable level of unwanted fringe shifts. In accordance with Equation 2.8, the replica of a signal wave is expressed as:

$$A_s \approx a_o \exp(\Psi_o) = a_o \exp\{i[\varepsilon (x, y) + \varphi_o]\} \tag{3.1}$$

Here, for simplicity, it is supposed that the aberrations caused by the hologram imperfections φ_h are negligibly small. A similar expression manifests a comparison wave at the output of a scheme of reconstruction:

$$A_c \approx a_o \exp(\Psi_c) = a_o \exp\{i[\varphi_c]\} \tag{3.2}$$

The distortions φ_o of reconstruction will be cancelled due to the interference with the comparison wave carrying the same aberrations: $\varphi_o = \varphi_c$; hence the resulting fringe shift will depend only on $\varepsilon (x, y)$.

It is interesting to note that Mach–Zehnder holographic interferometry is also able to cancel aberrations which notably exceed the visually unresolved fringe shift and are responsible for the quality of an interference background pattern (see, for example, Reference [25]). As a matter of fact, some experiments performed by the author (including those described in Section 3.1) show that distortions of reconstructed waves may exceed one dozen of the diagnostic wavelengths (see also Reference [8]). Thus holographic interferometry, owing to its differential character, makes it possible to perform accurate measurements of thin phase objects in spite of a large level of optical aberrations in schemes of recording holograms.

3.2.2   Sensitivity and Accuracy of Interference Measurements

The sensitivity S of any measuring technique is defined as the ratio of the changes in the recorded signal that is created at the instrument's output to the change of the parameter being measured.[26] In accordance with this definition, the sensitivity of an interference method means the ratio of the deformation of the wave front from an interferogram $\Delta\Phi_m$ to the deformation of the wave front which passed by the irradiated object $\Delta\Phi_i$ (actual change of the phase):

$$S = \Delta\Phi_m \ / \ \Delta\Phi_i \qquad\qquad (3.3)$$

In classic interferometry, sensitivity of the measurements has the same value: $S = 1$. As to the sensitivity of holographic interference measurements, this could be changed due to enhancing the deformation of the front of a signal wave in comparison with the wave front which passed by the irradiated object. This procedure can be done during the reconstruction of a signal hologram.

Let us assume that the maximum measured fringe shift caused by the phase object is only one quarter of a fringe: $\Delta N = \frac{1}{4}$; the accuracy A of the measurements can be evaluated as[27]

$$A = \frac{\Delta N_v}{\Delta N} \, 100\% = 40\% \qquad\qquad (3.4)$$

Thus it is impossible to perform interference measurements with better accuracy (smaller A) as long as the ΔN has the fixed value $\frac{1}{4}$, which is very important for numerical evaluation of the phase object under test. The only possible way to improve the accuracy of interference measurements of a weak phase object is by enlarging useful fringe shifts $\Delta N(x,y)$ on the reconstructed interferogram. In this case the sensitivity of interference measurements also will be improved and enhanced. Thus enlarging the deformation of the wave front of a signal wave is the only method for improving sensitivity and accuracy of interference measurements.

3.2.3 Traditional Methods of Enhancing the Sensitivity of Interference Measurements

It should be noted that the sensitivity increase has rather wide-reaching implications. First of all, it extends the range of applications of holographic interferometry to cases when a weak phase object was not previously observed visually. Second, weak phase objects cause small fringe shifts and give rise to maximum measurement errors due to insufficient accuracy. This fact is extremely important for reducing the interference data. An accurate calculation would be much better if the sensitivity is enhanced.

There are numerous examples in the literature of interference measurements with enhanced sensitivity; some of them are examined in Reference [8]. It is worth noting that the authors of References [28–30] have taken advantage of the nonlinearity of photo emulsions for the purpose of amplification of wave front deformations (another way of improving the sensitivity) of a reconstructed signal wave. Unfortunately some of the approaches are not able to completely compensate for simultaneously enlarging phase errors. It is easy to show that the phase distortions φ_o that accompany the useful phase changes $\varepsilon(x, y)$ are increased by the same factor during the procedure of amplification. Therefore, it is useless to increase front deformations of the signal wave without simultaneous compensation for the interferometer's phase imperfections (aberrations).

One of the first approaches to the compensation of aberrations is considered in References [31, 32], where, for the purpose of an improvement of the sensitivity, it is suggested recording secondary image holograms in symmetric diffraction orders. The authors showed that original imperfections of the order of ~5λ do not exceed λ/5 for a 14-fold increase in sensitivity.[32] The interested reader will also find in this article some useful references.

In aerospace engineering there are a few applicable approaches in which methods of amplification of useful phase information are discussed. These methods in particular are described in detail in Reference [33]. In Reference [34] the method of multi-pass holographic interferometry[35] was used to record low air density holograms in hypersonic flow at the High Enthalpy Facility F4 at ONERA. In the case of finite-width-fringe patterns, the fringe shift ΔN = 3 was achieved (with 6 passes through the shock field) in the R5Ch Wind Tunnel. The contrast of obtained flow (signal) holograms with each pass becomes increasingly worse, and the optical scheme of the holographic interferometer becomes less stable due to a large-scale reference arm. The method is subject to the same shortcomings mentioned in References [33, 34].

To improve the accuracy of recorded data, a phase-shifting interferometric (PSI) method was used in a conventional double-plate holographic reconstruction system;[36] the method achieved fractional fringe shift capability. PSI is a technique in which known phase shifts are introduced into the reference arm of an interferometer. The phase map is then calculated from the images recorded at each phase shift. This technique has an accuracy that is at least one order of magnitude better than the standard two-beam interferometry technique. At the same time, it should be noted that the PSI technique is quite cumbersome. Moreover, the technique cannot be directly applied if density changes across the shock wave are quite small, which means, that the fringe shift will also be small.[37] The method needs to be refined for this situation.

A very interesting approach to resonant holographic interferometry was suggested in References [38, 39] and has been developed for hypersonic gas dynamic flows in Reference [40]. Holographic interferometry measures the phase changes experienced by a propagating coherent laser beam due to changes in the real part of the index of refraction of the medium. The index of refraction changes dramatically in the vicinity of a resonant absorption feature. Thus, by tuning the laser line near the resonant feature, the enhancement of sensitivity of interference measurements can be achieved in comparison with the sensitivity for off-resonant interferometry. The maximum sensitivity can be achieved when $\lambda - \lambda_0 = \pm\Delta\lambda/2$, where λ is the wavelength of a laser line, λ_0 is the wavelength of an absorption line, and $\Delta\lambda$ is the width of the absorption line. The applicability of the technique is restricted by the species oscillating strength of air components and relatively short coherence length of tunable dye lasers.

3.2.4   Enhancement of Sensitivity by Amplification of Wave Front Deformations

Equation (3.3) definitively shows how to enhance the sensitivity of interference measurements, S, if we know how to increase the measured fringe shift ΔN of weak or small-sized phase objects. If interference data is processed visually, then the only chance to enhance the sensitivity of interference measurements is to increase the measured fringe shift from reconstructed interferograms.

There is one opportunity to increase the maximum measured fringe shift cardinally, but it exists only in holographic interferometry (dual hologram technique) owing to the canceling of the output distortions φ_0 in a reconstructed interferogram. From an experimental point of view, the approach is based on the enlargement of deformations of a signal's wave front; this enlargement is also called the *amplification of wave front deformation*. If, for example, it is possible to enlarge the deformation by a factor of four, the sensitivity of the measurements will be enlarged four times and the new accuracy in the above cited example would be of 10%.

As was described above, holographic interferometry demonstrates the character of a differential technique, just as it makes possible the successful study of the objects with phase changes which are remarkably smaller than $2\pi/10$. In order to properly study such phase objects by means of amplification of the wave front deformations, it is worthwhile to make use of the method of rerecording holograms suggested in Reference [31], and discussed in greater detail in Reference [41]. It seems that this promising and versatile method is a rather good fit for the procedure of enhancing sensitivity of interference measurements and can be widely recommended for practical applications. The technique of rerecording (rewriting) holograms produces amplification of wave front deformations and performs holographic subtraction of the unwanted phase information recorded on holograms so that the resulting aberrations in the interference pattern are canceled.

A signal hologram, being nonlinearly recorded during the run of a facility, serves as the first master hologram for rerecording the phase information existing in its first or higher diffraction orders. It is clear that, in this case, unwanted phase distortions will also be rerecorded on the secondary hologram. A comparison hologram is subjected to the same rerecording procedure.

It is relevant to remember that the master and consistent secondary (and so on) comparison holograms represent excellent "optical elements" in order that the "ideal" interference pattern will be reconstructed from the corresponding pair of signal and comparison Mach–Zehnder holograms. The interference subtraction is realized in the middle of reconstruction of the final pair of holograms in a dual hologram holder. The amplified distortions are the same for both reconstructed waves and will cancel each other out. By virtue of that, the resulting interference picture with the useful amplified phase changes will be free of optical aberrations.

3.2.5 Method of Rerecording Holograms

It is an experimentally fact that the majority of photographic films demonstrate nonlinear behavior of the curve $[\tau - H]$. In Mach–Zehnder holographic interferometry this nonlinearity is a rather useful feature, as it facilitates the amplification of phase information recorded on holograms. As was shown in Section 2.1, the amplitude transmission coefficient τ nonlinearly depends on exposure and can be described by Equation (2.4). In general, the amplitude transmission coefficient can be expressed as:

$$\tau(x, y) = \sum_{n=0}^{N} c_n \left\{ \exp\left[i(n\Delta\Psi) \right] + \exp\left[-i(n\Delta\Psi) \right] \right\} \tag{3.5}$$

Here c_n are some constants depending on conditions of the experiment and characteristics of the exposed photographic film. If, at the reconstruction stage, a hologram is illuminated by the reference wave $A_R \approx \exp[i(-2\pi fx)]$, then the waves reconstructed from the hologram in $2N + 1$ diffraction orders $A_{rec} = \tau \times A_R$ are described by the expression:

$$A_{rec} = \sum_{n=0}^{N} c_n \left\{ \exp\left[in\left(\varepsilon - 2\pi fx + \varphi_o\right)\right] + \exp\left[-in\left(\varepsilon - 2\pi fx + \varphi_o\right)\right]\right\} \quad (3.6)$$

Equation (3.6) describes a series of harmonics, propagating behind nonlinearly recorded holograms. Each harmonic represents one diffraction order with the phase difference, which is a positive or negative integer of the phase changes in the first order $\varepsilon(x, y)$.[14,15]

At normal incidence of the reconstructing reference beam on a negative, the hologram generates N positive and N negative orders (plus zero order). It follows from the grating equation:

$$\left[\sin\left(\delta\right) + \sin(i)\right] = \frac{n\lambda}{d} \quad (3.7)$$

where δ is the diffraction angle, $i = 0$ is the incident angle, n is the order of diffraction, λ is the wavelength, and d is the width of the holographic fringe.

If Mach–Zehnder signal and comparison holograms were recorded in the light of a helium–neon laser ($\lambda = 632.8$ nm) with the offset angle $\alpha = 2$ deg ($d \approx 0.018$ mm; $f \approx 55$ mm^{-1}), they theoretically generate 28 positive and 28 negative orders of diffraction ($N = 28$). The maximum diffraction angle is approximately equal to 78 deg. It is interesting to compare the results with a high spatial carrier frequency hologram. For example, if the offset angle is 15 deg, $f = 409$ mm^{-1}. There are $N = 3$ positive and 3 negative orders, with the maximum diffraction angle approximately equal to 51 deg. Equation (3.7) can be rewritten as: $\sin(\delta) = n \sin(\alpha)$, where α is the offset angle of the hologram. Thus, N can be defined from

$$N = \text{int}\left[\frac{1}{\sin(\alpha)}\right] \quad (3.8)$$

Any Mach–Zehnder signal hologram of a low spatial carrier frequency can be used to amplify information on the phase object recorded on it. For an amplitude plane grating, an intensity of the wave reconstructed in the positive first order does not exceed $^1/_{16}$ of the intensity of the reconstructing wave. As a rule, the experimentally obtained diffraction efficiency does not exceed 4%; in higher orders the intensity of reconstructed waves is lower yet. From this point of view, it would be more promising to use a phase grating instead of the amplitude one because of the ability to generate higher intensities in higher diffraction orders. The interested reader is referred to Reference [42], where a lot of useful information on phase holograms can be found. Unfortunately, there is one serious limitation regarding phase holograms: a relatively low signal-to-noise ratio (SNR) in higher orders.

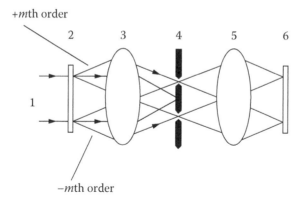

FIGURE 3.8
Optical scheme for rerecording holograms: (1) reconstructing wave; (2) master hologram; (3,5) collimator; (4) two-hole stop diaphragm; (6) secondary hologram.

The appropriate optical scheme for rerecording holograms is shown in Figure 3.8. A collimated reconstructing beam (1) illuminates a primary (master) hologram (2). Diffraction orders of only $\pm m$th orders are spatially selected by a two-hole stop diaphragm (4) in the back focal plane of an objective lens (3); zero and other orders are blocked out. (As a rule only first orders are used, because the intensity of higher orders quickly dies). The objective lenses (3, 5) comprise a telescope system, where the primary and secondary holograms are located in optically conjugated planes. The secondary signal hologram (6) is recorded by interference of $\pm m$th diffraction orders from the master hologram: $+m$th order wave is expressed as $\sim\exp[im(\varepsilon - 2\pi fx + \varphi_o)]$, and $-m$th order wave as $\sim\exp[-im(\varepsilon - 2\pi fx + \varphi_o)]$. The amplitude transmittance coefficient of the signal hologram in accordance with Equation 3.6 can be described by the following expression:

$$\tau_o(x,y) = \sum_{m=0}^{M} c_m \left\{ \exp\left[im(\Delta\Psi_o) \right] + \exp\left[-im(\Delta\Psi_o) \right] \right\} \tag{3.9}$$

The spatial carrier frequency of the secondary signal hologram is $2mf$, where f is the spatial frequency of the master signal hologram.

The same procedure is applied in order to record a secondary comparison (no flow) hologram: $+m$th order wave is $\sim\exp[im(-2\pi fx+\varphi_o)]$; $-m$th is $\sim\exp[-im(-2\pi fx + \varphi_o)]$.

$$\tau_c(x,y) = \sum_{m=0}^{M} c_m \left\{ \exp\left[i(m\Delta\Psi_o) \right] + \exp\left[-i(m\Delta\Psi_o) \right] \right\} \tag{3.10}$$

The spatial carrier frequency of the secondary comparison hologram is also $2mf$. Thus the secondary hologram is recorded with the spatial frequency amplified by a factor of $2m$ in comparison with the master hologram. For the number of the diffraction orders generated behind the secondary holograms, the value N is determined in accordance with Equation (3.8).

Thereafter the secondary signal and comparison holograms are spatially superposed in a dual hologram holder and illuminated by the reconstructing wave of unity amplitude, $A_R \approx \exp[im2(-2\pi fx)]$. This gives rise to generation of two reconstructed waves in the positive first order: $\sim\exp[2m(\varepsilon - 2\pi fx + \varphi_o)]$ and $\sim\exp[2m(-2\pi fx + \varphi_o)]$.

The resulting interferometric picture represents an infinite-width-fringe pattern, which can be expressed as:

$$I(x, y) \approx 1 + \cos[2m \times \varepsilon(x, y)] \tag{3.11}$$

If a series of finite-width-fringe interference pictures is required, the experimenter should follow the procedures described above.

Thus, deformations of a signal wave front are amplified by a factor of $2m$ in comparison with the master hologram. The aberrations φ_o of the interferometer's optics are also amplified by a factor of $2m$ in both holograms and are cancelled during the reconstruction. If the procedure is repeated q times, the sensitivity can be enhanced by a factor of $2mq$. It should be noted that, from a practical point of view, signal and comparison master holograms can be used for rerecording secondary holograms in the first symmetric diffraction orders, without taking advantage of nonlinear features of holograms. In this case, the phase changes will be amplified by a factor of 2 only, but the procedure may be repeated.

In connection with the procedure of rerecording holograms as described above, a challenging problem arises concerning the resolving power limitations of a photographic material. Indeed, the spatial frequency of final secondary holograms is $2mqf$ and may easily exceed the resolving power of the photo emulsion. If the master hologram has the spatial carrier frequency of ~25 mm^{-1}, the secondary hologram reconstructed in the third order requires a photo material with a resolving power larger than $2 \times 3 \times 25$ mm^{-1} = 150 mm^{-1}. In addition, the lens (3) in Figure 3.8 could vignette the reconstructed diffraction orders. To avoid that, master and secondary holograms can be reconstructed and recorded in special schemes with small holographing angles, which are analyzed in Reference [41].

The other limitation of the technique is a necessarily high accuracy of the superposition in a dual hologram holder. This requirement is increasingly difficult due to increasing sensitivity. Besides, for appreciable enhancements, the master holograms should have fewer aberrations. In conclusion, it should be noted that the described process may be performed successfully with double exposure holograms.

3.3   Applications of Mach–Zehnder Dual Hologram Interferometry with Enhanced Sensitivity

3.3.1   Suspended Droplets Vaporizing in Still Air

The experimental approach is discussed and described in References [22, 43]. The conventional dual hologram technique makes it possible to study weak phase objects which give rise to the fringe shifts of the order of ~ $1/20$. For vaporizing acetone droplets, with the maximum typical optical path length of the order of $\lambda/4$, this means a ~20% accuracy. It became apparent that in order to achieve, for example, 5% accuracy for such thin phase objects, the maximum measurable fringe shifts must be at least of the order of ~λ.

In order to improve the sensitivity of conventional interference analysis, the method of the amplification of wave front deformations of reconstructed signal waves was successfully applied. It follows from the previous analysis (Section 3.2) that useful phase changes in the *m*th diffraction order are *m* times stronger than the actual shift caused by the vapors of volatile liquids over a droplet. Droplet and no-droplet secondary holograms, obtained by using ±2nd orders from master holograms, were spatially superposed in a dual hologram holder and illuminated by the replica of a reconstructing wave, $A_R \approx \exp[-4i(-2\pi fx)]$. The intensity distribution in the resulting interference pattern had the form:

$$I \approx 1 + \cos[4\varepsilon(x, y) + 2\pi\omega y] \qquad (3.12)$$

where $\varepsilon(x, y)$ is the actual phase change caused by a vapor–air mixture, and ω is the spatial frequency of the background interference pattern (see Figure 3.9[b]). Note that the reconstructed interferograms in Figure 3.9 were obtained from holograms having vertical systems of carrier fringes of the same frequency, and in addition, no-droplet holograms were recorded without the fiber. Equation (3.12) describes an interferogram with the sensitivity increased by a factor of four, and with compensation-for-aberrations of the recording optical system.

It should be emphasized that the compensation-for-aberrations feature is the principal point for testing of such weak phase objects as air–vapor mixtures over droplets suspended on plastic fibers. For the first time,[43] the dual hologram approach with enhanced sensitivity was applied to study the dynamics of droplet vaporization of three types of liquids: acetone, n-pentane, and diethyl ether. Fringe shifts of ~1/80 for small size (< 0.5 mm) droplets with a low surface temperature were measured. Some results of the interference measurements are presented in Figure 3.10, where the radial distributions of the vapor densities are plotted versus the dimensionless radius of a droplet for all three liquids. The molecule vapor densities very close to the surface of a droplet are ~1.7×10^{18}, ~1.5×10^{18}, and ~0.6×10^{18} cm^{-3} for

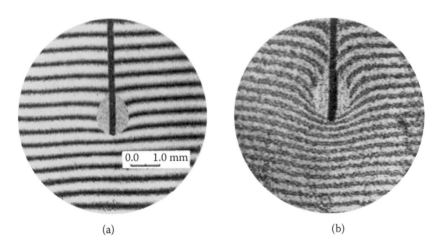

(a)   (b)

FIGURE 3.9
Reconstructed interferograms of vaporizing n-pentane droplets. Finite-width-fringe interferograms: (a) reconstructed from two master holograms; (b) reconstructed from two secondary holograms. Sensitivity was increased by a factor of four.

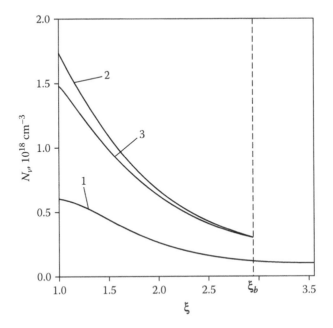

FIGURE 3.10
Radial distribution of vapor density versus dimensionless variable $\zeta = r/a$. (1) Acetone droplets with radius ≈ 760 µm; (2) n-pentane droplets with radius ≈ 680 µ; (3) diethyl ether droplets with radius ≈ 680 µm.

diethyl ether, n-pentane, and acetone, respectively. It is seen that n-pentane droplet vaporization dynamics are similar to diethyl ether.

3.3.2 Visualization of Acoustic Waves in Dye Solution

The experimental situation and the setup for pulsed holographic interferometry of acoustic waves generated in an ethanol dye are described in References [23, 24]. Dual hologram interference analysis with enhanced sensitivity of interference measurements was applied by means of the rerecording hologram technique. The second harmonic of a heating ruby laser was used to pump a dye laser with parameters $\lambda_d = 0.58$ µm, $E_d = 0.5$ mJ, and coherent length of ~1 cm. This light was used for testing the acoustic waves in a radial direction. The interference picture reconstructed from master signal and comparison holograms is presented in Figure 3.11(a). Analysis of interference data, owing to the enlarged sensitivity, showed that the change of the refractive index of the liquid Δn in the radiation-heated region is negative and has the value of the order of $\Delta n = -5 \times 10^{-3}$. In this case $(\partial n / \partial T)_p = -2 \times 10^{-5} \ T^{-1}$. In the compressed region of the acoustic wave, Δn reaches its maximum value of approximately -10^{-3} near the wall of the cell, and the magnitude of this value decreases by a factor of two at a distance of 1 mm from the glass wall, while the width of the compression region of the wave increases from 0.15 to 0.3 mm. The velocity of the cylindrical acoustical wave was estimated as 1.5×10^3 m $\times$ s^{-1} (the speed of sound in ethanol is ~1.16 km/s). A typical interferogram reconstructed for master and comparison secondary holograms obtained by the procedure of rerecording holograms is presented in Figure 3.11(b).

3.3.3 Testing of Supersonic Air Micro Jet Impinging Spherical Target

The Mach–Zehnder interferometer was used for making low carrier frequency (20 to 50 mm^{-1}) holograms on APX 25 Agfapan (100 mm^{-1}) 35-mm

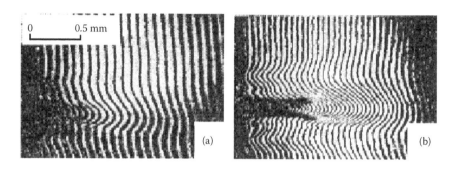

FIGURE 3.11
Interferograms reconstructed from (a) master holograms, and (b) secondary holograms with enhancement sensitivity factor of four. Time delay is 200 ns.

films.[3,4] Typical holographing angles were about one degree. Small angles have been chosen in order to work with photographic films having an ordinary resolving power. An argon ion laser with wavelength 514.5 nm was used as the source of coherent light, and the correspondent spatial frequency of holograms was ~34 mm[-1]. The maximal field of view was about 8 × 12 mm[2]. The spatial resolution of the recording scheme is ~0.02 mm. The infinite- and finite-width-fringe interferograms reconstructed from the master flow and no-flow holograms are demonstrated in Figure 3.12(a,c). The secondary holograms were recorded on Agfa-Gevaert photocopying film (300 mm[-1]). The secondary signal and comparison holograms were placed in a dual hologram holder and were used to generate the resulting interference patterns imaged in Figure 3.12(b,d). The interferograms shown in Figure 3.12(b,d) make it possible to characterize an air jet more precisely. The interferogram in Figure 3.12(d) was used in order to calculate radial distribution of air density in the jet, taking into account the axisymmetric form of the phase object. In this case, the data reduction was performed by the simplest ring method with constant density inside a ring.[8] The actual distribution of the density changes are smaller than calculated by a factor of four: $(\rho/\rho_\infty)_{act} = (\rho/\rho_\infty)_{cal}/4$. The results of calculations are presented in Figure 3.12(e), where

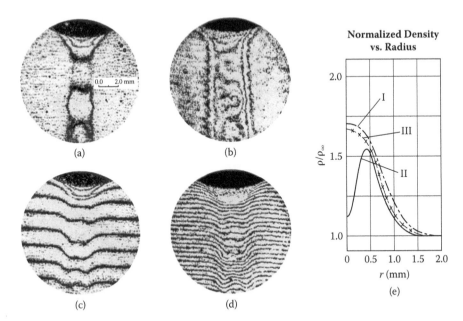

(a) (b) (c) (d)

Normalized Density vs. Radius

ρ/ρ_∞

(e)

FIGURE 3.12
Interference measurements of density in supersonic air micro jet with sensitivity enhancement. (a,c) Interferograms reconstructed from master holograms; (b,d) interferograms reconstructed from secondary holograms with sensitivity enhanced by a factor of four. Carrier frequencies: master holograms: f ~ 34 mm[-1]. (e) Normalized air density distribution versus radius in a supersonic air micro jet. Distances from sphere: (I) 0.9; (II) 2.45; (III) 4.0 mm.

the three smoothed curves represent calculated radial distributions of the density of air in a supersonic jet at three different distances from the spherical target. These three curves clearly show the periodic nature of the "cell" structure with successive shock and expansion waves along the flow axis, affecting maxima and minima of the density distribution. The data were compared with published table data. The divergence does not exceed ~10% for Mach number $M = 1.15$.

In Reference [4] the radial distributions of air density of a supersonic micro jet were confirmed by interference measurements using dual hologram shearing interferometry technique analyzed in Section 2.2.3.

The discussed approach of enlarged sensitivity of interference measurements has been demonstrated by analyzing the density field of a supersonic air micro jet, acoustic waves in a dye solution, and concentrated air–vapor mixtures over volatile liquid droplets. It has been shown that the method makes it possible to appreciably enhance the sensitivity of standard interference measurements with compensating optical aberrations. These remarkable features lead to more accurate phase measurements in the case of weak phase objects with only a fraction of the interference fringe shift.

Conclusion to Section I

Mach–Zehnder reference beam holographic interferometry is the simplest approach to optical holographic interferometry. Unlike standard methods, it does not required high-resolving-power holographic photo materials. Mach–Zehnder holograms could be recorded on wide types of standard photographic films having resolving powers 60 to 100 mm^{-1}. At the same time, the Mach–Zehnder approach guarantees applications of all the techniques developed for optical holographic interferometry. The advantages of this approach in comparison with its classic counterpart include: (1) relatively low optical quality of elements of the interferometer; (2) wide spectrum of coherent methods of studying signal waves; (3) unique opportunity of enhancement of the sensitivity of interference measurements of optically thin phase objects, which are difficult to investigate in frames of the classic technique; and the (4) opportunity of reconstruction for signal waves with canceling of some types of optical aberrations of the scheme of the holographic interferometer.

The only limitations are the impossibility of measuring very fine details of the phase objects and the necessity of using methods of spatial filtering during the process of reconstruction of Mach–Zehnder holograms.

The author hopes that the Mach–Zehnder optical reference beam holographic interferometry approach will become more and more popular in schools, universities, and companies' optical laboratories due to its simplicity and the small investments required to use this method.

Section II

Mach–Zehnder Optical Shearing Holographic Interferometry of Compressible Flows

Introduction to Section II

In Section I, phase objects having a field of visualization that does not exceed the aperture of a beam-splitting cube are studied by using methods of Mach–Zehnder reference beam holographic interferometry. At the same time, phase objects in aerodynamic applications as a rule have a large aperture, sometimes exceeding 500×500 mm^2 for testing sections of super- and hypersonic wind and shock tunnels. In that case, the author offers a slightly different test approach: Mach–Zehnder shearing holographic interferometry. This holographic approach has two principal advantages in comparison with the reference beam technique: (1) large apertures of the field of visualization (FOV) and (2) high acoustical stability of the optical setup. The last circumstance is important because practically all powerful aerodynamic facilities are sources of strong acoustical disturbances.

In Section II, methods and special techniques of Mach–Zehnder optical shearing holographic interferometry are considered. Indeed, focused image Mach–Zehnder holograms might not be only reference type, but shearing also. In Section II, the diffraction shearing holographic interferometer intended for studying flow fields in aerodynamic facilities is carefully analyzed. Distinctive characteristics of the proposed scheme for recording the Mach–Zehnder shearing signal and comparison holograms are discussed. The setup is acoustically stable, which allows recording of signal holograms even in a noisy supersonic wind tunnel facility by using continuous wave (CW) lasers. The scheme is applicable for different powerful gas-dynamics and similar facilities that have a large aperture of the test sections.

Different methods of Mach–Zehnder shearing optical holographic interferometry are applied during investigation of compressible shock flows in a supersonic wind tunnel: double exposure, dual hologram, and real-time techniques. Special techniques of regulated sensitivity of the interference method are also analyzed. A few examples of applying Mach–Zehnder shearing holographic interferometry are presented: shock flows, subsonic turbulent jets, and convection laminar and turbulent flows.

4

Mach–Zehnder Shearing Holographic Interferometry and Its Applications in Aerodynamics

4.1 Problems of Application of Interference Techniques in Gas Dynamic Research

Optical interferometry, with a laser as the light source, has been used as a diagnostic tool in fluid dynamic research for longer than 40 years. However, the gas dynamic flows that could be addressed with such an approach were restricted due to numerical engineering problems, mainly because of the extreme sensitivity of reference beam optical schemes to vibrations during the run of an aerodynamic facility, and excessive expanses involved in large-aperture interference systems. Thus, the two challenging problems that must be addressed before designing interferometric optical schemes for large-scale gas dynamic facilities are mechanical stability and large apertures of optical elements of the interferometer. From the interference point of view, the problem is that mechanical vibrations affecting the interferometer's optical elements lead to a low contrast of recorded flow (signal) interferograms/holograms during the run. The low contrast prevents the recording of interference pictures of satisfactory quality, or even makes it impossible.

This problem was partially alleviated by the application of pulsed laser systems, acoustically isolated platforms, and specially developed mechanically stable interferometer schemes. By using a pulsed laser system, vibrations of the optical elements of a setup tend to insignificance because the exposure duration is extremely short: ~5 to 50 ns for Q-switched laser systems. On the other hand, random displacement of the optical and mechanical elements during the run could still lead to spurious interferometric fringes in the middle of reconstruction, or even to abrupt decrease of the contrast of a recorded flow (signal) hologram if a continuous wave (CW) laser is used as the coherent light source. Moreover, the reference and object arms in classic and ordinary holographic schemes are particularly sensitive to acoustic disturbances; hence they are spatially separated and independently subjected to mechanical vibrations. In addition, if schemes are complicated, having very long optical lengths of object and reference arms with the reference beam passing under or over a wind tunnel via an optical trench, all these circumstances complicate alignment procedures, making them time consuming.

Taking into account these circumstances, numerous interferometers, in which an object wave, or both reference and object waves pass through the viewing windows of a test section or an evacuated chamber, are more attractive for performing interference measurements in comparison with reference beam schemes with spatially separated arms. The attractiveness of such schemes is that they could be designed on the basis of existing Schlieren devices and could have lower vibration sensitivity and temperature characteristics.

One of these schemes with both beams traversing an object was reported in Reference [44], but it may be applied to quite coarse phase objects and is more sensitive in one of the two perpendicular directions, in the vertical plane (x, y). In Reference [45], the double pass conventional Schlieren setup is converted to an interferometer by replacing a knife-edge with a suitable spatial filter. Only infinite-fringe interferograms can be recorded in real-time mode.

A similar interferometer's concept was described in Reference [46], where the diagnostic beam is divided into two components to produce object and reference waves after passing the test field; this is the class of point-diffraction interferometers (PDI). The PDI technique uses the light passing through the pinhole in an optical filter[47] or in a photographic plate emulsion[48] located at the back focal plane of the second Schlieren mirror. A focused laser beam powerful enough to burn a hole in a photo material or a coating filter, with no flow in a wind tunnel, generates the pinhole. Because of the spatial characteristics, the light passing through the pinhole loses all the phase information introduced by the flow; therefore a reference wave is created in the light beam passing beyond the pinhole. This reference wave subsequently interferes with the light that was transmitted around the pinhole through the photographic plate or the filter, producing interference fringe in real time at the image plane. The disadvantages of the technique are the following: the inability of the spatial filters to withstand even intermediate levels of the pulsed laser energy, and only an infinite-fringe mode interferogram recording.

More flexible and operational is the technique proposed in Reference [49]. The setup is the same as for conventional Schlieren schemes, excluding the receiver section. At the receiver, the test field beam is split into two in the same manner as in common Mach–Zehnder interferometers. One of the two beams is spatially filtered through a pinhole diaphragm to get the reference beam. The object beam retains wave distortion information produced by the phase object under study. The beams recombine in the image plane, giving a real-time interference picture. Nevertheless, optical components of the receiver section must be made very compact and rigidly mounted to minimize mechanical vibrations. In addition, experiments show that optical component imperfections make impossible effective focusing and filtering of the test beam to produce the reference beam due to aberrations.

The same obstacle regarding noncompensated aberrations introduced by two spherical mirrors of the Schlieren section prevents proper filtering of the test beam in the optical scheme with diffraction grating, which was suggested in Reference [50]. It is supposed that the reference beam is introduced by spatial filtering of one of the diffraction orders.

Thus, the main disadvantages of the interference techniques discussed above are: (1) a limited possibility for recording or reconstructing interferograms in finite-width-fringe mode; (2) the necessary use of expensive long-coherence-length pulsed laser systems; (3) the necessary use of auxiliary, high-quality expensive optics because aberrations are not compensated in conventional spherical mirror Schlieren schemes; (4) low acoustical stability caused by spatially separated reference and object beams.

Numerous experiments on gas dynamic facilities at the Faculty of Aerospace Engineering, Technion, including a $M = 4.0$ supersonic wind tunnel,[51] have shown that the above described optical scheme constructions are not applicable in the case of using CW argon ion or HeNe lasers as the coherent light sources due to remarkable acoustical vibrations and noncompensated aberrations in the test section of an interferometer. On the other hand, a series of experiments at the supersonic wind tunnel made it possible to develop and design an optimized holographic lateral shearing interference scheme that meets strong acoustical and aberration requirements. Moreover, suggested optical principles of the design effectively facilitate a high mechanical stability of the optical scheme in comparison with any reference beam holographic interferometer's scheme.

Conventional shearing interferometry[52–54] also possesses a majority of the shortcomings inherent to the reference beam interference schemes. Literature analysis shows that holographic shearing interferometers[55,56] were successfully applied to study some phase objects; however they are not adapted to wind tunnels and other large-aperture, acoustically unstable gas-dynamic facilities.

In the next section, a novel dual hologram shearing interference technique[57] and corresponding interferometer, which can be easily constructed on the basis of an existing conventional Schlieren system, is described. Its mechanical stability allows signal holograms to be recorded during the run of a facility by using a continuous wave laser light source, which could also be used in applicable real-time holographic techniques.

Recorded Mach–Zehnder holograms are placed in a dual hologram holder for subsequent processing. The reconstructed interferograms are obtained with compensation for optical aberrations. Moreover, the dual hologram technique makes it possible to get the background interference fringes with any arbitrary orientation and spacing. The shearing interference technique provides a simple way for displaying the first and the second difference quotients of the phase changes and allows differential interferograms to be obtained. The double exposure technique has been suggested in Reference [58].

4.2 Concept of Mach–Zehnder Dual Hologram Shearing Interferometry for Gas Dynamic Applications

Formally, the approach described here[57,59,60] is in some respects related to the double exposure method of shearing interferometry[58] and the dual hologram interferometry technique.[8] Due to a minimal spatial separation between the two sheared diagnostic waves, the scheme imaged in Figure 4.1 can be successfully used for wind tunnel, as well as shock tunnel or other powerful facility testing, because its mechanical and temperature characteristics are close to those of conventional Schlieren devices.

The scheme is applicable for any large-aperture facility that has a high level of mechanical instability, and possesses all important characteristics of classic shearing interferometers. In addition, that it has the tolerance to Schlieren optical quality windows in the test section and Schlieren mirrors as wide-aperture optical elements.

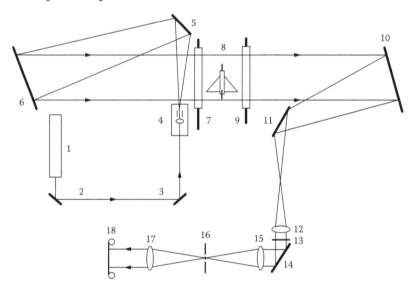

FIGURE 4.1

Scheme of diffraction lateral shearing holographic interferometer for testing in wind tunneld and other wide-aperture facilities for aerospace research. (1) Argon ion laser; (2, 3) flat mirrors; (4) spatial filter; (5) flat mirror; (6) spherical Schlieren mirror; (7) viewing window; (8) model; (9) viewing window; (10) spherical Schlieren mirror; (11) flat mirror; (12) collimating objective lens; (13) phase grating; (14) flat mirror; (15, 17) collimator (telescope); (16) two-pinhole diaphragm; (18) photographic film.

4.2.1 Interferometer

The proposed holographic lateral shearing diffraction interferometer shown schematically in Figure 4.1 (compare with the scheme suggested in Reference [58]) is designed on the base of a conventional single-pass Z-type Schlieren device with large-aperture (320-mm diameter) spherical mirrors (6,10). A narrow-pencil argon ion laser (1) beam is cleaned by a spatial filter (4). Beyond the optical filter, due to a microscopic lens, the expanding spherical wave is collimated by the first Schlieren mirror (6). The system micro-lens and the mirror (6) present a collimator; moreover, the focal plane of the mirror coincides with the pinhole of the spatial filter. A collimated diagnostic beam passes through two Schlieren quality windows (7, 9) of the test section and a flow field over the model (8).

The wave carrying optical information on the compressible flow field is received by the second Schlieren mirror (10) and refocused to the receiver section. A lens (12) is used to collimate the spherical beam from the mirror (10) and illuminate a phase grating (13). The grating is located in the plane, which is optically conjugated to the vertical plane, symmetrical to the model, i.e., the flow field under study is sharply focused by the collimator (10, 12) on the grating (13). Let us consider the design of the receiver section in more detail.

A closer look at the receiver is presented in Figure 4.2. A collimator (15, 17) images the phase grating on a Mach–Zehnder hologram (18) (the diameter of which is ~25 mm) with magnification M_g corresponding to the ratio of the diameters of a circular spot on the phase grating (13) to the diameter of the collimating beam illuminating the hologram. The lens (15) refocuses the collimated beams or, in other words, the diffraction orders (only the

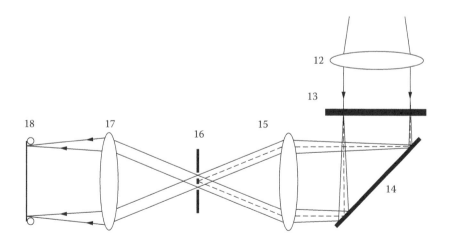

FIGURE 4.2
A closer look at the receiver section of the holographic shearing diffraction interferometer. (12) Collimating objective lens; (13) phase grating; (14) flat mirror; (15, 17) collimator; (16) two pinhole diaphragm; (18) photographic film.

lowest three orders are shown in Figure 4.2), in the plane of a two-pinhole diaphragm (16) in order to select only the plus and minus first orders. Zero and higher diffraction orders are blocked out.

The photo film (18) is thus exposed to two nonshifted waves passed by the object under test; herewith the spatial frequency of the hologram f is determined by the spatial frequency of the phase grating f_g and the coefficient of magnification $M_g \times f = 2f_g/(M_g)$. As long as the grating and the hologram exist in optically conjugated planes, this frequency does not depend upon the wavelength λ of the laser (1). It should be emphasized that the spatial frequency of the grating should be small enough so that the spatial frequency of the output Mach–Zehnder hologram does not exceed the resolving power of photographic film (18) (in reality MTF should be larger than 0.5).

The maximal spatial resolution of the technique is determined by the coefficient of total magnification $M_t =$ (diameter of hologram/diameter of diagnostic wave) and the spatial frequency of the shearing hologram f_h. In accordance with Section I, the smallest detail which could be seen in a reconstructed interferogram has the spatial frequency of $\sim 2/f_h$ mm^{-1}. Taking into account the coefficient of magnification M_t, the theoretical resolution of the optical system is of the order of $\sim M_t \times (2/f_h)$ mm. The experimentally determined value is at least three times larger. It seems that this discrepancy might be explained by two factors: (1) the smallest phase details of the flow field cannot be recorded on the signal hologram due to aberrations of the Schlieren mirror (10), and (2) the low optical quality of the windows in the test section.

Translating the phase grating along the z-axis, i.e., in the direction normal to itself, one may control a lateral shear between the two collimated waves on the hologram (18). The grating generates laterally sheared waves in the first orders with a zero optical path difference, thus eliminating problems due to the failure to maintain the equal optical path requirements occurring in other shearing interferometers. The grating performs the combined functions of the beam-splitter and the shearing element. The grating's spatial frequency, magnification, and the magnitude of translations, Δz, determine the shear distance between the diagnostic waves. By observing the doubled model's image in the plane of the hologram and translating the grating, proper shear distance Δx can be picked up visually. The real shear in the FOV is computed as $\Delta x/M_t$. Note that the optical scheme only allows for production of a lateral shear in the direction perpendicular to the carrier fringes of a hologram (vertical holographic fringes in Figure 4.2).

4.2.2 Holographic Recording

There are two pairs of sheared waves of the signal and comparison waves, i.e., during and before (or after) the run, which should be recorded on two separate holograms: (1) the signal (or flow) carrying phase information on the flow field under test, and (2) the comparison (no flow), which carries an information on aberrations of the optical scheme. It should be noted that the same aberrations are coded on the signal hologram. Due to the collimating lens (17) and the diaphragm (16), these two shifted first orders appear on the photo film for recording Mach–Zehnder shearing holograms.

Let us suppose for simplicity that the grating is imaged on the hologram with unity magnification. A pair of the positive and negative first orders of the signal wave, diffracted from the phase grating and striking the photo film, can be described as:

$$U_{S/+} = \exp\{i[\varepsilon_+(x, y) + \phi_+(x, y) - 2\pi f_g x]\},$$

$$U_{S,-} = \exp\{i[\varepsilon_-(x, y) + \phi_-(x, y) + 2\pi f_g x]\} \tag{4.1}$$

and the pair of comparison waves are described as:

$$U_{c+} = \exp\{i[\phi_+(x, y) - 2\pi f_g x]\},$$

$$U_{c-} = \exp\{i[\phi_-(x, y) + 2\pi f_g x]\} \tag{4.2}$$

Here, ε is the phase change caused by the inhomogeneity being investigated, ϕ is the integral phase distortions of the wave front caused by aberrations of the interferometer's optical elements and the grating, and f_g is the spatial frequency of the phase grating.

A signal hologram formed by laterally sheared waves (Equation [4.1]) with the shear Δx would have an amplitude transmittance that is proportional to:

$$\tau_s(x, y) \approx 2 + \exp[i(\Delta\varepsilon + \Delta\phi - 4\pi f_g x)] + \exp[-i(\Delta\varepsilon + \Delta\phi - 4\pi f_g x)] \tag{4.3}$$

where $\Delta\varepsilon = \varepsilon_+ - \varepsilon_- = \varepsilon(x + \Delta x/2, y) - \varepsilon(x - \Delta x/2, y)$, and $\Delta\phi = \phi_+ - \phi_- = \phi(x + \Delta x/2, y) - \phi(x - \Delta x/2, y)$ take place for the horizontal lateral shear Δx.

A comparison hologram is formed with the same lateral shear Δx. Its amplitude transmittance is proportional to:

$$\tau_c(x, y) \approx 2 + \exp[i(\Delta\phi - 4\pi f_g x)] + \exp[-i(\Delta\phi - 4\pi f_g x)] \tag{4.4}$$

4.2.3   Hologram Reconstruction

Let the signal and the comparison holograms, being spatially superposed in a dual hologram holder, be reconstructed by a plane wave $U_{rec} \approx \exp[i(4\pi f_g x)]$. The only two waves, $\exp[i(\Delta\varepsilon + \Delta\phi)]$ and $\exp[i(\Delta\phi)]$, will propagate in the direction of the positive first order and generate an infinite-fringe interference pattern, $I(x, y) \sim 1 + \cos[\Delta\varepsilon(x, y)]$. The latter relation shows that the reconstructed interferogram does not depend on the phase distortions $\Delta\phi(x, y)$ that appear due to some types of aberrations. Note that both holograms have the vertical systems of carrier fringes.

A horizontal finite-width-fringe interferometric pattern and its spatial frequency could be produced and controlled by a dual hologram holder as follows. After being adjusted by the holder to form an infinite-width-fringe interferogram (both holograms have the same spatial carrier frequency), signal and comparison holograms should be rotated at a certain angle with respect to each other in order to obtain a series of finite-width-fringe interferograms with horizontal fringes. Note that the sign of the background fringe gradient in the vertical direction depends on the direction of rotation.

A shearing interferogram with background fringes in one direction, only gives the information on wave front in the direction perpendicular to the fringes. Thus the to get full information on the wave front deformation, one requires two shearing interferograms with shear in the two orthogonal directions.[61] This can be accomplished by rotating the phase grating at 90 degrees over the z-axis that guarantees the horizontal system of background holographic fringes. In this case, the two-pinhole diaphragm should also be rotated at 90 degrees.

In shearing interferometry, it is customary to correlate the magnitude of a shear and the value of the phase gradient, so that smaller shears are used for the objects that have steeper gradients.[62] For small lateral shear, the interference fringes represent certain values of the first difference quotient $d\varepsilon/dx$:

$$\frac{\varepsilon\left(x+\Delta x/2, y\right)-\varepsilon\left(x-\Delta x/2, y\right)}{\Delta x} = \frac{d\varepsilon}{dx}\left(x, y\right)+\vartheta_1\Delta x\frac{d^2\varepsilon\left(x+\vartheta_2\Delta x, y\right)}{\Delta x^2} \quad (4.5)$$

where $-1 < \vartheta_1, \vartheta_2 < 1$.

In addition, in the case of stationary phenomena, the second phase derivative can be displayed as well. If two signal holograms with lateral shears Δx_1 and $\Delta x_1 + \Delta x_2$ are spatially superposed in a dual hologram holder, there is the chance to record $\Delta^2\varepsilon(x, y)/(\Delta x_1\Delta x_2)$. Here, the resulting intensity of the infinite fringe interference pattern will be:

$$I(x, y) \approx 1+ \cos\left[(\Delta\varepsilon)_2 - (\Delta\varepsilon)_1\right]$$

where

$$\left(\Delta\varepsilon\right)_2 -\left(\Delta\varepsilon\right)_1 =\left[\varepsilon\left(x+\frac{\Delta x_1}{2}+\frac{\Delta x_2}{2}, y\right)-\varepsilon\left(x-\frac{\Delta x_1}{2}+\frac{\Delta x_2}{2}, y\right)\right]$$

$$-\left[\varepsilon\left(x+\frac{\Delta x_1}{2}, y\right)-\varepsilon\left(x-\frac{\Delta x_1}{2}, y\right)\right]\approx \frac{d^2\varepsilon\left(x, y\right)}{dx^2}\Delta x_1\Delta x_2$$

$$(4.6)$$

If a nonstationary phenomenon has high speed and good reproducibility, it might be useful to apply a pulsed laser system for recording the two flow holograms in two different runs.

4.2.4 Verification of Proposed Experimental Setup

To verify the proposed approach and to demonstrate its ability in the area of gas dynamic research, a series of diagnostic experiments (see the references and the descriptions of the experiments in Section 4.3, Chapter 5, and Section 6.1) have been performed with diverse phase objects.

As the sources of coherent radiation, an argon ion laser with the output power of the order of ~50 mW at wavelengths 488.0 and 514.5 nm and a Q-switched ruby laser pulse ~400 mJ at the wavelength of 694.3 nm and a duration of $\tau_{1/2}$ = 30 ns have been used. Mach–Zehnder holograms with low spatial frequency holographic fringes that varied from 50 to 160 lines × mm^{-1} were recorded on Agfa photo film (300 mm^{-1}) conventionally used for micro-filming, Fuji SS-DX 35-mm photo film (100 mm^{-1}), and holographic photo plates PFG-01, which are sensitive for 694.3 nm radiation.

The main optical element of the receiver section is the diffraction grating (13). Ordinary overexposed amplitude gratings were bleached in order to convert them to higher diffraction efficiency phase gratings. The diffraction efficiencies were measured as ~20% and 4% for the phase and amplitude gratings, respectively.

Unlike Mach–Zehnder reference beam interferometry, in shearing interferometry there is an issue of overlapping images of a model, which obstructs the procedure of adjusting the signal and comparison holo-grams in a dual hologram holder. In order to avoid this effect, the com-parison (no-flow) hologram should be exposed without the model in a test section before (or after) the run of a wind tunnel. Otherwise, only the exact overlapping can be achieved for the infinite-width-fringe interfero-gram mode.

4.2.5 Schlieren and Shadow Diagnostics of Wave Front Reconstructed from a Shearing Signal Mach–Zehnder Hologram

The wave reconstructed from a signal shearing hologram can be success-fully studied by applying holographic shadow and Schlieren techniques, as was already described in Section 2.2. Indeed, if two replicas of a signal wave, $\exp\{i[\varepsilon_+ + \Phi_+ - 2\pi f_g x)]\}$ and $\exp\{i[\varepsilon_- + \Phi_- + 2\pi f_g x)]\}$, are used to record the hologram, its amplitude transmittance can be presented in the form of Equation (4.3.)

If a negative is illuminated by the reference wave $\exp[i(4\pi f_g x)]$, the signal wave $\exp[i(\Delta\varepsilon + \Delta\Phi)]$ will be reconstructed in the first order. For the purpose of reconstruction and study of the useful deformation $\Delta\varepsilon(x, y)$ caused by the phase changes of the object under investigation, the optical schemes imaged in Figure 2.3(b) can be used. The phase distortions $\Delta\Phi$ caused by optical aberrations of the recording should be small enough to prevent errors that can be obstructive to numerical treatment of a Schlieren or shadow picture.

4.3 Applications of Mach–Zehnder Holographic Shearing Interferometry to Supersonic Wind Tunnel Testing

4.3.1 Dual Hologram Shearing Interferometry of the Shock Flow Around the Blunt-Nose Cone Model

A 400×500 mm^2 supersonic wind tunnel at the Aerodynamic Research Center[51] (the Technion) was used to produce supersonic flow fields to be studied. An axisymmetric cylindrical model with blunted nose at zero angle of attack was used to generate the shock wave flow field variations. Tests were run at Mach numbers $M = 2.0$–2.5. Mach–Zehnder shearing holograms were recorded by using the holographic diffraction shearing interferometer shown in Figure 4.1. In each experiment stagnation and static pressures were determined. Typical reconstructed interferograms of a supersonic flow field around the blunt nose cone model with diameter of 40 mm and half cone angle of 30 deg are shown in Figure 4.3. The various figures are but a few examples that show the flexibility of the proposed technique.

Sensitivity of the technique is proportional to the shear spacing, s, which is demonstrated in Figure 4.3. The three interferograms in Figure 4.3(a) and Figure 4.3(b,c) were reconstructed by using two different signal holograms recorded with the shears of 0.1 and 0.2 mm, respectively. The interferograms (Figure 4.3[b,c]) were reconstructed by using the same comparison hologram. Because the density of the interference fringes in the area of the normal part of a bowed shock is too high to be resolved, the fringe pattern has to be decided upon from the discontinuity at infinity, where the shock wave degenerates to a Mach wave with infinitesimal strength.

Note the region of very low contrast (indicated by the arrow) in the interferograms with the larger shear (Figure 4.3[b,c]). For shears of the order of $\Delta x = \Delta y$ at 0.1 mm or smaller, a low-contrast fringe region does not occur. One may reasonably assume that significant aerodynamic effects, turbulence and/or oscillations of the flow in this region, dramatically affect the contrast of the reconstructed interference patterns. The degradation of the contrast will be more pronounced if shears are large enough relative to the scale of turbulence or the amplitude of oscillations. Thus, this rather simplistic interferometric approach makes it possible to see specific turbulent regions of flows and estimate the scale of turbulence (unsteadiness) in such regions by varying the magnitudes of shears.

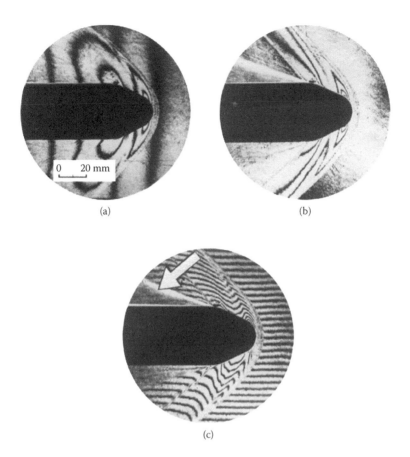

FIGURE 4.3
Reconstructed shearing interferograms of shock wave flows around a blunt nose cone model.
(a, c) Finite-width-fringe interferograms; (b) infinite-width-fringe interferogram. The shears
(in the field of vision): (a) $\Delta y = 0.1$ mm; (b, c) $\Delta x = 0.2$ mm. Mach number is $M = 2.0$.

The shearing technique enables one to see and measure density gradients
only in the direction of the beam shears; however, this direction may be varied for the next run. Interferograms of mutually perpendicular beam shears
must be used to determine the density gradient vectors over the entire field
of view.[63] This means that the design of a shearing interferometer should
enable shears to be produced easily in mutually perpendicular directions;
the two pictures of the flow field under study should be recorded with the
same magnification. These requirements could be a complicated problem for
a classic optical scheme. The measurements of the density gradient vectors
over the entire field of view could be easily realized by using the described
interferometer. Shears in the perpendicular direction could be arranged by
rotating of the phase grating (and the diaphragm) by 90 deg. Thus the interferometer's design provides shears in mutually perpendicular directions

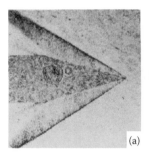 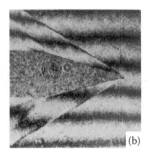 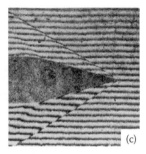

FIGURE 4.4

Reconstructed shearing interferograms of shock wave flow around sharp nose cone model. (a) Infinite-width-fringe; (b, c) finite-width-fringe interferograms. Shears (in the field of vision) are parallel to free stream flow velocity vector: (a, b, c) $\Delta x = 0.1$ mm. The diameter of the model is 25 mm; half angle is 15 deg. Mach number is $M = 2.2$.

only by means of rotating the grating (13) and the stop diaphragm (16) without any change of the design.

4.3.2   Dual Hologram Shearing Interferometry of the Shock Flow Around the Sharp-Nose Cone Model

In Figure 4.4 a shock flow around a sharp nose cone model can be observed. All three interferograms were reconstructed from one pair of signal (flow) and comparison (no-flow) Mach–Zehnder holograms with the same spatial frequency of holographic carrier fringes that are located in the vertical direction.

4.3.3   Pulsed Double-Exposure Holographic Shearing Interferometry of Shock Flows in $M = 4.0$ Wind Tunnel

The diffraction shearing interferometer has been also successfully applied to wind tunnel testing by using a Q-switched pulsed laser system as the coherent light source. The optical scheme of these holographic experiments is similar to that shown in Figure 4.1. The first difference is that the CW argon ion laser was replaced with a pulsed ruby laser system; the CW laser was used for alignment. The Q-switched pulsed laser system is capable of producing ~400 mJ energy pulses of a short pulse duration $\tau_{1/2}$ ~30 ns to "freeze" the shock flow under study. The coherence length is longer than ~100 cm (line width is ~10^{-2} cm^{-1}) at the wavelength 694.3 nm. The powerful pulsed laser system makes it impossible to use the spatial filter (4) because using even relatively low energy levels of focusing the laser beam leads to the initiation of a laser spark in the vicinity of the filter, which may easily destroy the spatial filter's pinhole. To avoid this laser spark effect, a negative uncoated glass lens (singlet) was used to expand the laser beam in order to overfill an aperture of the first spherical Schlieren mirror (6). To obtain a parallel beam it is necessary to control the coincidence of the negative lens and the Schlieren mirror (6) foci. The final change was made to the collimating objective (17), which was replaced by a large-aperture long-focal-length objective lens to expose photographic plates with apertures larger than the size of spot of diameter of 25 mm (up to 90 mm).

Comparative experiments have been performed to understand the difference between reconstructed interferograms of shock flows with the same characteristics obtained with the CW argon ion and the pulsed ruby lasers.

It is obvious that the extremely short pulse of the ruby laser should "freeze" shock flow and free stream nonuniformities so that they can be visualized on interferograms and compared with the one-millisecond "long pulse" of the CW laser. The comparison experiments showed that the shortest exposure time at which holograms were recorded by using the CW argon ion laser, of the order of 1 ms, is really very long in comparison with characteristic times of gas-dynamic disturbance/turbulence. Thus, gas dynamic parameters of the object under test are averaged. A series of shearing interferograms reconstructed from the holograms recorded by using the pulsed ruby system are presented in Figure 4.5.

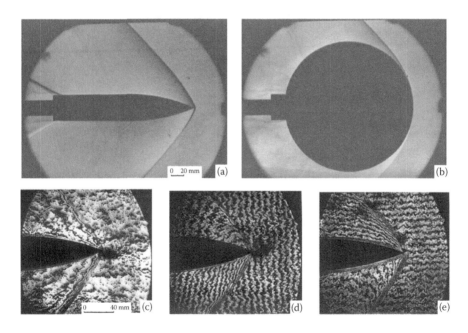

FIGURE 4.5
Using a pulsed ruby laser for visualizing a shock flow over "flying plate." (a,b) White light source Schlieren photographs; (c–e) reconstructed shearing interferograms; (c) infinite- (d,e) finite-fringe-width interferograms. Shears (in the field of vision): (c, d) $\Delta y = 0.2$ mm; (e) $\Delta x = 0.3$ mm. Mach number $M = 1.7$. Unlike white light Schlieren photographs, turbulent nonuniformities could be observed in the case of using the pulsed ruby laser system, which "freezes" the flow field nonuniformities on the reconstructed shearing interferograms.

5

Mach–Zehnder Shearing Optical
Holographic Interferometry
with Regulated Sensitivity

5.1 Applications of Mach–Zehnder Shearing Holographic Interferometry in Aerospace Engineering Research

All interferograms which will be discussed in this section have been reconstructed from Mach–Zehnder holograms, which were recorded by using the interferometer imaged in Figure 4.1.

5.1.1 Turbulent Heated Air Jet from Heat Gun

The subsonic turbulent heated air jet generated by a heat gun is one of the interesting phase objects for studying temperature turbulence.[64] Figure 5.1(a,c) represents a series of interferograms reconstructed from Mach–Zehnder holograms with the shears varied from 0.1 to 0.3 mm. All holograms were recorded by using a continuous wave (CW) argon ion laser at the wavelength of 488.0 nm. This series demonstrates the effect of enlargement of interference fringe shifts depending on the values of shears, s. The region with strong temperature turbulence can be observed in the right-hand side of the pictures, where a very low contrast of interference fringes is directly connected to a low contrast of the corresponding area in the recorded holograms.

The ability of the technique for displaying a second derivative of the index of refraction is demonstrated in Figure 5.1(d). The displayed second derivative was obtained by simultaneous reconstruction of two signal holograms with the shears of 0.1 mm.

In all figures, the contrast of interference fringes appear to be degenerated by means of turbulent fluctuations during a long time exposure of ~10 ms. The attempt to resolve these turbulent fluctuations was undertaken by applying the minimal possible exposure time of 1 ms, but it failed, because the characteristic time of the turbulent flow is shorter. Such fluctuations, it

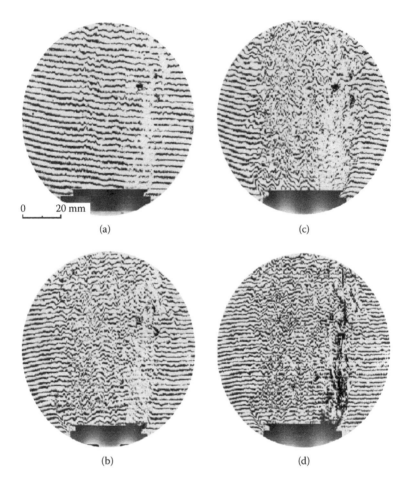

FIGURE 5.1

Reconstructed shearing interferograms of turbulent heated air jet from a heat gun. (a–d) Finite-width-fringe interferograms. Shears are perpendicular to the averaged jet velocity vector: (a) $\Delta x = 0.1$ mm; (b) $\Delta x = 0.2$ mm; (c) $\Delta x = 0.3$ mm. Visualization of second derivative: (d) $\Delta x_1 = \Delta x_2 = 0.1$ mm. Diameter of the heat gun's nozzle is 41 mm.

seems, could be "frozen" by applying a much shorter duration pulse from a pulsed laser as the source of a coherent light.

The reconstructed interferograms in Figure 5.2 represent transient turbulent (a,b) and laminar (c,d) convection regimes. The signal holograms were recorded 5 and 15 seconds after switching off the heat gun, respectively. The interferograms shown in Figure 5.2(a,b) were reconstructed from the pairs of signal and comparison holograms, which had the same spatial carrier frequencies. To obtain the finite-width-fringe interferograms in Figure 5.2(c,d), one signal and one comparison hologram were rotated in a hologram holder at different angles (holographic fringes are horizontal). Signs of the gradient

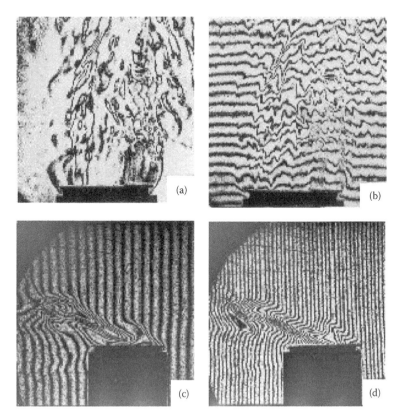

FIGURE 5.2
Reconstructed Mach–Zehnder shearing interferograms of transient turbulent (a) and (b), and laminar convection (c) and (d) regimes. (a) Infinite- (b,c,d) and finite-width-fringe interferograms. (a, b) 5 sec; (c, d) 15 sec after heat gun's switching off. Shears (in the field of vision): (a, b) $\Delta x = 0.12$ mm; (c, d) $\Delta y = 0.12$ mm.

of the background fringe patterns are opposite because of the opposite directions of the rotation.

All the interferograms demonstrate the ability of the technique to reconstruct infinite- and finite-width-fringe interference patterns having only one signal hologram, as well as to change the direction of the gradient of a background fringe pattern.

5.1.2 Conical Volatile Liquid Spray

A series of experiments was conducted with a Burke–Schumann combustor to study axial symmetry and characteristics of liquid volatile sprays generated by a commercial ultrasonic atomizer.[65] Note that fuel droplets are injected as a narrow conical spray. The fuel flow rate was changed during the experiments, as was the type of the volatile liquid: *n*-hexane and

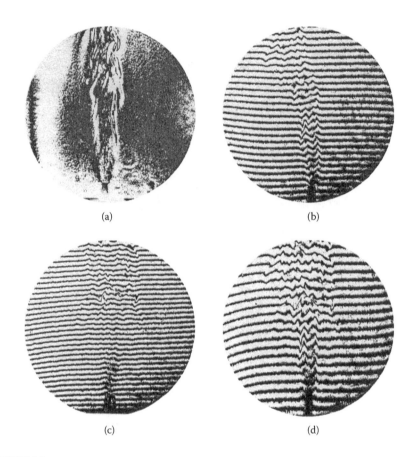

(a)   (b)

(c)   (d)

FIGURE 5.3
Reconstructed shearing interferograms of conical volatile liquid spray in air. (a) Infinite-width-fringe interferogram; (b, c, d) finite-width-fringe interferograms. (a) n-Hexane; flow rate is 0.31 liter × hour^{-1}. (b, c, d) n-Heptane; flow rates are (b) 0.51 and (c) 0.31 liter × hour^{-1}. Shears (in the field of vision) are: $\Delta x = 0.12$ mm. Diameter of the field of view is ~200 mm.

n-heptane. The representative reconstructed Mach–Zehnder shearing interferograms are shown in Figure 5.3. It should be remarked that all the holograms were recorded under circumstances of high mechanical instability of
the optical scheme, using the light of a CW argon ion laser with wavelength
of 488.0 nm.

The infinite-fringe interferogram shown in Figure 5.3(a) allows the boundaries of the n-hexane spray and its form to be determined. The interferogram
illustrates the appearance of a stray bright fringe caused by high mechanical
instability of the optical scheme. The comparison of the two interferograms
in Figure 5.3(b) and (c) shows that the process of vaporization at the lower
flow rate (0.31 liter × hour^{-1}) goes more quickly and effectively because of
larger fringe shifts and lower contrast of the interference fringes in the vicinity of a nozzle. Figure 5.3(c) indicates a higher concentration of droplets. The

reconstructed interferogram in Figure 5.3(d) displays the phase difference between the two liquid spray regimes presented in Figure 5.3(b) and (c). This differential interferogram was obtained by reconstructing two corresponding signal holograms simultaneously.

5.2 Mach–Zehnder Dual Hologram Shearing Interference Technique with Enhanced Sensitivity for Wind Tunnel Testing

5.2.1 Concept of the Technique

Dual hologram shearing interferometry with enhanced sensitivity was proposed for visualizing and measuring the density gradients of compressible flows in a supersonic wind tunnel.[66–68] The technique is especially useful for a low free-stream density or a strong turbulent flow. The technique is characterized with tolerance to both vibration disturbances characteristic of many wind tunnel facilities, and to Schlieren optical quality of the test section windows used for viewing flow phenomena and spherical large-aperture Schlieren mirrors. The proposed optical method is related, in some respects, to the dual hologram shearing interference technique suggested in Reference [57] and to the reference beam holographic technique with enhanced sensitivity, described in detail in Section 3.2. This method is able to perform phase gradient measurements in the areas where diagnostic methods of conventional shearing interference analysis have failed or are not easily applied.

When a wave probes a phase object, shearing interferometry offers the opportunity to determine gradients of the wave front deformed in the direction of the beam shear, $s = \Delta x$: $k\partial\Phi/\partial x = k\alpha_a$, where k is the wave number, Φ is the optical path length difference, and α_a is the angle of arrival. The sensitivity of the conventional technique can be changed proportionally to a shear spacing (the magnitude of a shear), s. If this value is a small quantity, it is permitted to write the fringe equation in the following form:

$$2\pi N(x,y) = ks\frac{\partial\Phi}{\partial x} = ks\int_{z_1}^{z_1}\frac{\partial}{\partial x}n(x,y,z)dz = (ks)K\int_{z_1}^{z_1}\frac{\partial}{\partial x}\rho(x,y,z)dz \quad (5.1)$$

where K is the Gladstone–Dale constant, and ρ is air density.

A significant turbulence or a high level of flow fluctuations, existing in some regions of the flow field under investigation, may dramatically affect the contrast of the interference pattern. Moreover, it was confirmed experimentally that the degradation of the contrast due to a subsequent enlargement of the shear, s, becomes even more pronounced when turbulence or unsteadiness exists. Taking into account the results obtained in Reference [64] for a Moiré deflectogram, one may show that the holographic carrier fringe contrast degradation is controlled by the phase fluctuations, $\varphi^* = k \times \Phi^*$, related to the angle of arrival fluctuations α_a^*: $s^2 <(\partial\varphi^*/\partial x)^2> = (k \times s)^2 <\alpha_a^{*2}(x, y, L)>$, where L is the propagation distance of the diagnostic light wave through the turbulent field. All the values are averaged with respect to the time of exposure. In this case, the averaged amplitude transmission coefficient of a signal hologram, $<\tau(x, y)>_S$ can be written in the form:

$$\langle\tau\rangle_s \approx \left[1-(ks)^2\left\langle\alpha_a^{*2}\right\rangle/2\right]\left\{2+\exp\left[i\left(\langle\Delta\varepsilon\rangle-4\pi\Omega x\right)\right]+\exp\left[-i\left(\langle\Delta\varepsilon\rangle-4\pi\Omega x\right)\right]\right\}$$

(5.2)

where Ω is the spatial frequency of the phase grating (13) in Figure 4.1.

The amplitude transmission coefficient for a comparison hologram can be written in the form: $\tau_s = 2 + \exp(-i4\pi\Omega x) + \exp(i4\pi\Omega x)$. In the +1st order will be the two reconstructed waves: $\exp(-i4\pi\Omega x)$ and $[1 - (ks)^2<\alpha_a^{*2}>/2]\exp[i(<\Delta\varepsilon> -i4\pi\Omega x)]$.

Finally, the intensity pattern of the corresponding reconstructed interferogram can be expressed as:

$$I(x, y) \approx 1 + [1 - (ks)^2 <\alpha_a^{*2}>/2] \times \cos(<\Delta\varepsilon>) \qquad (5.3)$$

The last equation shows that the degradation of the amplitude of the interferometric term depends on $(ks)^2<\alpha^{*2}(x,y,L)>/2$, which is proportional to s^2. Thus, it turns out that a subsequent enlargement of the shear, s, with the purpose of enhancing of the sensitivity, becomes less practical in the case of dramatic fringe contrast degradation due to the turbulence or unsteadiness.

Equation (5.1) shows that the sensitivity is dependent on the magnitude of a shear, s, solely in the case of continuously varying gradients of the density. In the case of infinite density gradients on a shock wave, Equation (5.1) cannot predict the fringe pattern. Only those pairs of conjugate rays—which have one ray passing behind, and the other passing in front of the shock surface—can contribute to the formation of the respective pattern. One derives that the relative fringe shift is:

$$2\pi N\left(x, y\right) = k\int_{z_1}^{z_1} \Delta n(x, y, z)dz \qquad (5.4)$$

with Δn being the index of refraction jump through the shock. Equation (5.4) has the same form as the respective equation for reference beam interferometry and clearly shows that the fringe shift on a shock wave does not depend upon a shear, s. In other words, the fringe shift on a shock cannot be changed while increasing the magnitude of a shear. However, the approach of rerecording holograms would make it possible to uniformly enhance the sensitivity of the phase gradient measurements for the whole flow field, including shocks. It will also be demonstrated below that novel approach enables performing phase gradient measurements in unsteady (or turbulent) regions where diagnostic methods of conventional shearing interference analysis are rarely applied.

The independently proposed methods of enhancing the sensitivity of measurements by optical processing of holographic lateral-shear interferograms were recently considered in References [68, 69].

5.2.2 Experimental Results

To verify the proposed approach and to demonstrate its abilities, a series of diagnostic experiments have been conducted at Mach number $M = 2$, with an axisymmetric flow field around the spherically blunted 30-deg half-angle cone model, and with the ratio of nose radius to base radius of 0.5. Base radius is 22.5 mm. An argon ion laser with the output power of the order of ~60 mW at the wavelength of 488.0 nm was used as the coherent radiation source. Holograms with the spatial carrier frequency varying from 55 to 80 lines × mm^{-1} were recorded on an Agfa–Gevaert 35-mm photo material, designed for microfilming. The resolving power of the photographic film is ~300 mm^{-1}, and the time exposure was ~1 ms.

The receiver module of the holographic lateral shearing interferometer with a phase grating is shown in Figure 4.2. Due to the grating's translation and the spatial filtering, each of the two diffracted first orders of the flow and no-flow waves appear on the respective hologram as shifted beams of the input diagnostic beam.

It is important that each hologram is recorded in the image plane of the objective lens (17), which images a flow field under test on the hologram (18). Note that a reconstructed interferogram correctly transfers a phase difference between the signal and comparison interfering beams only if a phase object is focused in the plane of the hologram.

Typical Mach–Zehnder reconstructed interferograms of the supersonic flow field over the blunt-nose cone model with two different shear spacings are demonstrated in Figure 5.4(a) and (b). The interferograms were reconstructed from two pairs of master flow and no-flow holograms recorded with different shears of 0.1 and 0.2 mm, respectively. As already noted, for shears of the order $s = 0.1$ mm (Figure 5.4[a]) or smaller, the low contrast fringe region does not occur. In the second interferogram with the doubled shear, $s = 0.2$ mm, it can be seen that the turbulent structure of the flow, i.e., the region of very low contrast (shown by the arrow) in the interferogram, is identified. It is clear that, for the enlarged shear spacing,

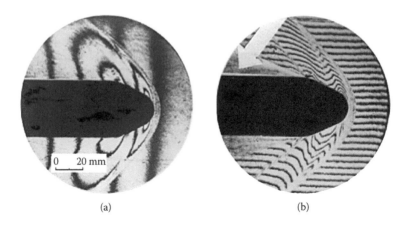

(a)   (b)

FIGURE 5.4
Shearing interferograms reconstructed from two pairs of flow and no-flow master holograms
with different shears. Shear spacing (in the field of vision) is perpendicular (a) and parallel to
free-stream velocity vector (b). (a) $s = \Delta y = 0.1$; (b) $s = \Delta x = 0.2$ mm. Arrow shows the turbulent
region.

a strong turbulence leads to the loss of interferometric information in this
region.

One may show that the proposed technique allows the sensitivity of interference measurements to be enhanced without any loss of the information in
strong turbulent regions. For that purpose, the flow and the no-flow holograms
with the shear spacing $s = 0.1$ mm were used as master holograms for generating rerecorded secondary holograms with increased wave front deformations.

First, secondary flow and no-flow holograms are recorded by using the
±1st diffraction orders reconstructed from the master holograms. Second, the
secondary holograms were used to record tertiary holograms in ±1st diffraction orders, which are selected by a stop diaphragm, while other orders are
blocked out. A dual hologram holder is used for the procedure of rerecording holograms, as well as at reconstructing stages, to control superposition
of the holograms by precisely varying their orientation with respect to each
other.

Figure 5.5(a) shows the interferogram with background fringes, which are
perpendicular to the velocity vector, reconstructed from master ($s = \Delta y = 0.1$
mm) holograms. Figure 5.5(b) and (c) demonstrate the interferograms that
were reconstructed from rerecorded secondary and tertiary flow and no-
flow holograms.

The latter two interferograms demonstrate the flexibility of the technique to enhance the sensitivity of interference analysis by a factor of 2
and 4, respectively. The capability of performing density gradient measurements with enhanced sensitivity in the area where turbulence or
unsteadiness prevents enhancing the sensitivity by enlarging the shear
spacing, is also demonstrated.

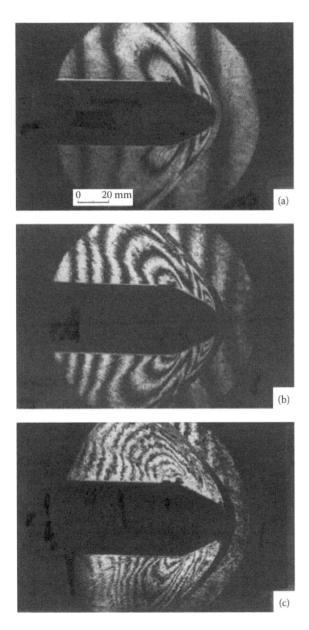

FIGURE 5.5
Reconstructed shearing interferograms with sensitivity enhanced by a factor of (a) 1, (b) 2, and (c) 4. Master holograms were recorded with shear spacing (in the field of vision) $\Delta y = 0.1$ mm. (a) Interferogram was reconstructed from master holograms; (b) interferogram was reconstructed from secondary holograms; (c) interferogram was reconstructed from tertiary holograms.

The recording signal (flow) and comparison (no-flow) conditions on separate holograms demonstrates some attractive capabilities in comparison with conventional shearing interference analysis. These capabilities include the ability to generate a series of reconstructed shearing interferograms with the same shear and different sensitivities by applying the rerecording process. The technique demonstrates the ability to perform density gradient measurements with enhanced sensitivity in the whole flow field, including shock flow regions, where a strong turbulence or unsteadiness efficiently degrades the contrast of interferometric fringes and prevents the enhancing of the sensitivity by enlarging a shear spacing. In other words unlike classic shearing interferometry, the technique is able to increase the sensitivity of phase gradient measurements uniformly for the whole flow field, including the shocks.

The technique is especially capable of measuring phase gradients in low free-stream density wind-tunnel flows owing to enhanced sensitivity and the compensation of optical imperfections. This is of importance especially in the case of Schlieren-grade optical quality of the tunnel windows, which were not designed for interference measurements.

It is possible that the above approach will become a useful tool for visualization and accurate mapping of the density gradients in strong turbulent gas dynamic or low free-stream density flows, where vibration disturbances from the wind tunnel facility may prevent the application of reference beam interferometric schemes.

5.3 Phase Measurements in a Supersonic Shock Flow with Reduced Sensitivity: Two-Wavelength Mach–Zehnder Holographic Shearing Interferometry

5.3.1 Theoretical Concept

Experimental efforts in the last two or three decades have shown that holographic reference beam interferometry is a powerful diagnostic tool for gas-dynamic studies, including wind tunnel testing. This technique has the potential of providing reliable optical information about a flow field, but the biggest difficulty has always been the retrieval of data from the interferograms.[70]

One of the specific measuring problems that occur in compressible aerodynamics is discontinuous data caused by a step change of the fringe shift, ΔN_{sh}, across a shock wave. This problem can be handled only if ΔN_{sh} can be accurately estimated[8] by introducing some complementary information. This information can be obtained from experimental data or from a computational fluid dynamics (CFD) experiment. In Reference [70], a method of avoiding ambiguity in the interpretation of interferograms near a shock is proposed. The method is based on the continuity of the double-exposure Schlieren method and holographic interferometry. In the CFD technique, the calculated errors of ΔN_{sh} may be appreciable owing to both a severe bow shock and a high sensitivity of interference measurements. In the case of shearing interferometry, the step change of the fringe shift ΔN_{sh} across a shock can be expressed by the following relation (compared with Equation [5.4]):

$$\Delta N_{sh}(x, y) = \frac{1}{\lambda} \int_{z_2}^{z_1} \Delta n(x, y, z) dz \qquad (5.5)$$

Here $\Delta n (x, y, z)$ is the step change of the refractive index, and λ is the wavelength of a diagnostic light.

In the case of continuously varying parameters in the flow field, if the amplitude of the shear s is a small quantity, one may write the fringe equation in the form, which is close to Equation (5.1):

$$\Delta N_{sh}(x, y) = \frac{s}{\lambda} \int_{z_1}^{z_2} \frac{\partial}{\partial x}\left[n(x, y, z)\right] dz \qquad (5.6)$$

This equation shows that the sensitivity of the interference shearing technique, in the case of a continuously varying refractive index gradient, can be changed with s. To illustrate the above reasoning, refer to Figure 5.4, where two reconstructed interferograms of the same flow field with different shears, $s = \Delta y = 0.1$ and $s = \Delta x = 0.2$ mm, are presented. On the other hand, the regulation of the sensitivity in the case of an infinite refractive index gradient on a shock wave front is not possible when s is changed.

A characteristic feature of shearing interferometers is the formation of the doubled image on the edge of solid models. The doubled image is projected into the direction of the shear, and the exact position of the model's edge is in the middle of the gray zone or doubled image. Because the density gradient through the shock is infinite, Equation (5.6) cannot predict the fringe pattern in this case. Only those pairs of conjugate rays—which have one ray passing behind, and the other passing in front of the shock surface—can contribute to the formation of the respective pattern[62] (see Equation [5.5]).

Thus, Equation (5.5), having the same form as the respective equation for reference beam interference technique, clearly shows that the step change of the fringe shift $\Delta n (x, y, z)$ does not depend upon the shear, s. On the contrary, it will be shown in this paragraph that the two-wavelength interference technique with reduced sensitivity is applicable even in the case of a shock wave.

An interesting approach to reference beam holographic interferometry with reduced sensitivity has been suggested and experimentally demonstrated in References [71–73]. The single-exposure two-wavelength, λ and Λ ($\Lambda > \lambda$), holographic technique has been applied to show that the sensitivity of an interference analysis can be remarkably decreased, with limits $0 < \mu_d < 1$, where μ_d is the coefficient of sensitivity decrease:

$$\mu_d = \frac{\lambda}{\lambda_{\text{eff}}} \tag{5.7}$$

where $\lambda_{\text{eff}} = [\lambda\Lambda/(\Lambda - \lambda)]$.

The purpose of this section is to report some preliminary experimental results obtained by applying a novel two-wavelength holographic shearing interference analysis for measuring the fringe shifts in shock wave flows, including the region of a shock itself.[74] The approach would be promising, if it were possible to reduce the maximal step fringe shift of the shock wave flow to a value that does not exceed unity. Then the step fringe shift could be measured directly from the interferogram without introducing some complementary information on the strength of a shock.

It will be shown that reconstructed shearing interferograms have fringe shifts smaller than one. Such an approach could allow interference data to be retrieved from the reconstructed interferogram without any complementary assumptions regarding the step change of ΔN across a shock, which is of special interest in the case of strong shocks in a wind tunnel and strong traveling shock waves.

5.3.2   Experimental Realization

The holographic variable shear interferometer, designed on the basis of a conventional single-pass Z-type Schlieren scheme, is shown in Figure 4.1. The shock wave flow around a model is illuminated twice (in the two different runs of a facility), by two collimated coherent light beams of wavelength λ and Λ from a CW argon ion laser.

To verify the proposed approach, a series of experiments was conducted at the supersonic wind tunnel with the test section 400×500 mm^2. In each experiment stagnant pressure, static pressure, and Mach number were recorded. As the source of coherent radiation, the argon ion laser with an output power of ~60 mW at wavelengths of 514.5 and 488.0 nm was used. A sharp-nose cone/cylinder model was chosen. The axisymmetric cylindrical model with a sharp nose, at zero angle of attack, served to generate the shock wave flow field variations. Tests were run at Mach numbers $M = 2.0$ and 2.2.

5.3.3   Dual Hologram Approach with Reduced Sensitivity

A signal hologram formed by laterally sheared waves having the wavelength λ with the shear Δx would have an amplitude transmittance which is proportional to (see Equation [4.3]): $\tau_s(\lambda) \approx 2 + \exp\{i[\Delta\varepsilon(\lambda) + \Delta\phi(\lambda) - 2\pi fx]\}$ $+ \exp\{-i[\Delta\varepsilon(\lambda) + \Delta\phi(\lambda) - 2\pi fx]\}$, where f is the carrier spatial frequency of the hologram. A signal hologram recorded by using the wavelength Λ with the shear Δx would have an amplitude transmittance that is proportional to: $\tau_s(\Lambda)$ $\approx 2 + \exp\{i[\Delta\varepsilon(\Lambda) + \Delta\phi(\Lambda) - 2\pi fx]\} + \exp\{-i[\Delta\varepsilon(\Lambda) + \Delta\phi(\Lambda) - 2\pi fx]\}$. Note that both holograms recorded in the light of different wavelengths have the same carrier spatial frequencies.

If recorded with different wavelengths λ and Λ, the two signal holograms are superimposed in a dual hologram holder and reconstructed simultaneously by a plane wave $\sim\exp[i \times (2\pi fx)]$ with the wavelength λ. Only two waves, $\exp[i(\Delta\varepsilon(\lambda) + \Delta\phi(\lambda)]$ and $\exp[i(\Delta\varepsilon(\Lambda) + \Delta\phi(\Lambda)]$, are propagating in the direction of the positive first order, where $\Delta\varepsilon$ describes the phase variation introduced by the flow field around the model, and $\Delta\phi$ is the phase distortion of the wave front caused by aberrations of the optical recording system, including the phase grating. The shearing interference pattern of an infinite-width-fringe mode is proportional to:

$$1 + \cos(\Delta\varepsilon_{red} + \Delta\phi_{red}) \qquad (5.8)$$

where $\Delta\varepsilon_{red} = \Delta\varepsilon(\lambda) - \Delta\varepsilon(\Lambda)$, $\Delta\phi_{red} = \Delta\phi(\lambda) - \Delta\phi(\Lambda)$. Because we can neglect the nondispersive aberration term $\Delta\phi_{red} \approx 0$ $[\phi(\lambda) \approx \phi(\Lambda)]$, the resulting shearing interference pattern can be written in the form:

$$I_{res} \approx 1 + \cos(\Delta\varepsilon_{red}) \qquad (5.9)$$

One may show that for continuously varying gas-dynamic parameters in a shock flow, as well for the step changes of the index of refraction in a shock front, the sensitivity of interference measurements is reduced by the factor of $\mu_d = \lambda/\lambda_{eff}$: $\Delta\varepsilon(\lambda)/\Delta\varepsilon_{red} = \mu_d$. It is assumed that the index of refraction is the nondispersive value. This expression, in other words, means that the interferogram obtained in the two wavelengths has sensitivity which is μ_d times smaller than the sensitivity of an ordinary interferogram recorded and reconstructed by the same wavelength λ. Taking into account that $\Lambda = 514.5$ and $\lambda = 448.0$ nm, the sensitivity will be reduced by the factor of 19.41, and $\lambda_{eff} = 9{,}475$ nm. It should be noted that the resulting interference pattern obtained with reduced sensitivity is practically free of the phase distortions caused by aberrations of the optical recording system and the phase grating, because they are cancelled.

5.3.4 Experimental Results

Two signal holograms, which have been recorded with different wavelengths of $\lambda = 448.0$ and $\Lambda = 514.5$ nm, were superposed in a dual hologram holder and illuminated by laser light with wavelength $\lambda = 448.0$ nm. Typical

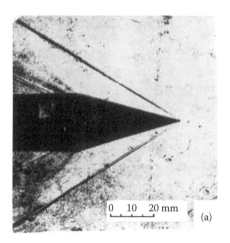 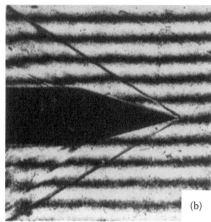

FIGURE 5.6

Lateral shearing interferograms with reduced sensitivity of shock flow around sharp-nose cone model. Interferograms are reconstructed from the two signal holograms, which were recorded in light of $\lambda = 448.0$ and $\Lambda = 514.5$ nm. (a) Infinite-width-fringe interferogram with reduced sensitivity; (b) finite-width-fringe interferogram with reduced sensitivity; $\mu_d = 19.41$. Mach number is $M = 2.2$. Cone half angle is 15 deg. Diameter of the cylinder is 25 mm. Shear (in the field of vision) is $\Delta x = 0.2$ mm.

reconstructed interferograms of a supersonic flow field around the sharp-nose cone model are presented in Figure 5.6. Both interferograms were reconstructed from the same pair of signal holograms. The infinite- and finite-width fringe interferograms obtained have reduced sensitivity, as can be definitively seen in Figure 5.6. It should be noted that the signal holograms were recorded with the different wavelengths and at the same time have the same spatial carrier frequencies.

A variety of methods can be used to find density gradient distribution around the axisymmetric sharp-nose cone model. The simplest method divides the flow region into a series of discrete annular elements of constant width and density gradient.[8,9] Unfortunately, one of the difficulties connected to the annular ring method is that it breaks down when step discontinuities of the density gradient occur.[75] An example of this is the region near a shock wave and especially near a strong bow shock.

The approach is capable of conducting simplified numerical analysis of the density gradient distribution in shock flows without any complimentary assumptions because the fringe shift, ΔN_{sh}, across a shock can be reduced to a value less than 1. This technique could be used to conduct simplified numerical analysis of the shock flows of very high densities, which are hard to study using conventional shearing methods. The technique, with reduced sensitivity and practically full compensation for optical aberrations, makes it possible to eliminate the difficulties of interference analysis of areas where huge step discontinuities in the density gradients occur.

6

Mach–Zehnder Real-Time Shearing Optical Holographic Interferometry

6.1 Real-Time Holographic Interference Shearing Technique for Wind Tunnel Testing

6.1.1 Concept of Design of Real-Time Holographic Shearing Interferometer

The method described in this section (see Reference [76]) is in some respects related to real-time reference beam holographic interferometry described in References [8, 9] (see Section 2.3.3). For experiments, the optical scheme presented in Figure 4.1 was used as the basis for designing an applicable setup that meets requirements of the real-time approach. The proposed optical scheme of the receiver is shown in Figure 6.1(a). The only principal addition to it in comparison with Figure 4.1 is the Mach–Zehnder comparison hologram (7) and some optical elements behind it.

The phase grating (3) is placed in the image plane of the flow field under study by means of a telescope (1, 2). The second telescopic system (4, 6) images the phase grating onto the comparison hologram (7). The stop diaphragm (5) selects only the first diffraction orders; others orders are blocked out.

After adjusting the optical scheme by using a pilot helium–neon laser (not shown in the figure), the phase grating (3) is translated along the optical axis in order to shear two waves of the diagnostic beam, which are generated in ±1st symmetrical orders behind the grating. The no-flow (comparison) hologram (7) is recorded in light of a Q-switched pulsed ruby laser (λ = 694.3 nm). The pilot helium–neon laser (λ = 632.8 nm, P = 10 mW) is used not only for the purpose of adjustment, but also in order to support working with the real-time regime interferometer.

The first stage of the experiment consists of recording the comparison hologram (7), which could be accurately and easily repositioned after the exposure. In the second stage, which is carried out in real time, the two selected first diffraction orders, or positive (+1) and negative (−1) signal waves, illuminate the comparison hologram (7) and generate the (+) and (−) zero orders comparison waves in the middle of the real-time regime. The "living" signal

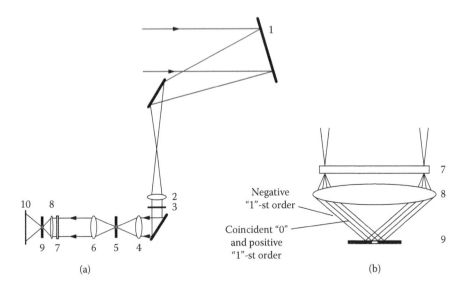

FIGURE 6.1

(a) Receiver module of real-time holographic shearing interferometer and (b) schematic zoomed layout of selecting diffraction orders behind comparison hologram. (1) Second Schlieren mirror; (2) collimating objective lens; (3) phase grating; (4, 6) collimator; (5) stop diaphragm; (7) comparison hologram; (8) objective lens; (9) stop diaphragm; (10) screen.

and reconstructed comparison waves in both zero orders generate two identical interference patterns of living fringe. Let us consider the appearance of the living fringe pattern in detail step by step.

6.1.2 Comparison Hologram Recording

As was already mentioned, the comparison hologram (7) is recorded on a photo plate (holographic photo material PFG-01M) by using a Q-switched pulsed ruby laser light with a long (~1 m) length of coherence. Let us suppose that the phase grating is imaged on the hologram with a magnification of unity. Due to the translation of the phase grating, both waves appear on the photo plate already being shifted ($\Delta x = s$):

$$U_{c'+} = \exp\{i[\varphi(x + s/2, y) - 2\pi \times f_g \times x)]\} \qquad (6.1)$$

$$U_{c'-} = \exp\{i[\varphi(x - s/2, y) - 2\pi \times f_g \times x)]\}$$

where φ is phase distortions of the wave front caused by aberrations of Schlieren-grade optical quality wind tunnel windows and relatively low-optical-quality elements, including the phase grating. The carrier frequency of the comparison hologram is defined as $f = 2f_g$, where f_g is the spatial frequency of the phase grating. The comparison hologram recorded on the photo plate has the coefficient of amplitude transmittance, which is proportional to:

$$\tau_c(x, y) \approx 2 + \exp[i(\Delta\varphi - 4\pi \times f_g \times x)] + \exp[-i(\Delta\varphi - 4\pi \times f_g \times x)] \qquad (6.2)$$

where $\Delta\varphi = \varphi(x + s/2, y) - \varphi(x - s/2, y) = \varphi_+ - \varphi_-$

6.1.3   Hologram Reconstruction and Interferometric Pattern Generation

A pilot He–Ne laser is used as the source of continuous wave (CW) coherent radiation. First, an accurate repositioning of the comparison hologram is required in order to minimize the quantity of extraneous fringes that could occur. Second, when the comparison hologram is repositioned in its original place, the two signal waves propagating in the first orders behind the phase grating illuminate the comparison hologram during the run of the supersonic wind tunnel (or other gas-dynamic facility). These sheared signal waves can be presented in the form:

$$U_{c+} = \exp\{i[\varepsilon(x + s/2, y) + \varphi(x + s/2, y) - 2\pi \times f_g \times x)]\} \qquad (6.3)$$

$$U_{c-} = \exp\{i[\varepsilon(x - s/2, y) + \varphi(x - s/2, y) + 2\pi \times f_g \times x)]\}$$

where ε is the phase change caused by the nonuniformity under investigation.

There are two pairs of waves traveling in the same direction (see Section 2.3.3). The first pair is $2\exp[i(\varepsilon_- + \varphi_-)]$ and $\exp[i(\varepsilon_+ + \varphi_+ - \Delta\varphi)]$. The resulting interference pattern in an infinite-width-fringe mode will be:

$$I_{inf} \approx 5 + 2 \exp[i(\Delta\varepsilon)] + 2 \exp[-i(\Delta\varepsilon)] = 5 + 4 \cos(\Delta\varepsilon) \qquad (6.4a)$$

The second pair $2\exp[i(\varepsilon_+ + \varphi_+)]$ and $\exp[i(\varepsilon_- + \varphi_- + \Delta\varphi)]$ will give the resulting interference pattern in the symmetric order, which is expressed as:

$$I_{inf} \approx 5 + 2 \exp[i(\Delta\varepsilon)] + 2 \exp[-i(\Delta\varepsilon)] = 5 + 4 \cos(\Delta\varepsilon) \qquad (6.4b)$$

The latter two equations show that both of the reconstructed "living fringe" patterns are free of phase distortions $\Delta\varphi$ caused by the integral effect due to the wind tunnel's test section windows and the phase grating. Thus phase aberrations responsible for a relatively low quality of optical elements in the scheme may be cancelled in the two coincident "zero-first" orders. Similar to the reference beam real-time holographic interferometry (see Section 2.3.3), Equations (6.4a) and (6.4b) illustrate that the contrast of fringes in the resulting interference pattern does not exceed ~80%.

A background finite-width-fringe pattern could be produced in the same way as described in Section 1.2.3. The vertical system of background interference pattern is expressed as:

$$I_{fin} \approx 5 + 4 \cos(\Delta\varepsilon + 2\pi \times \Omega \times y) \qquad (6.5a)$$

where Ω is the spatial frequency of horizontal background interference fringes. The vertical background system of interference fringes is described as:

$$I_{fin} \approx 5 + 4 \cos(\Delta\varepsilon + 2\pi \times \Omega \times x) \qquad (6.5b)$$

where Ω is the spatial frequency of background interference fringes.

The only two coincident zero-positive-first and two first-negative diffraction orders are shown schematically in Figure 6.2(b). Either of the two coincident zero-positive-first interference patterns can be selected by the diaphragm (9) and imaged on the screen (10) (or on the sensitive area of a CCD/CMOS sensor or photo film).

6.1.4   Experimental Results

The holographic shearing interferometer was designed in the Supersonic Wind Tunnel Laboratory (the Technion) to perform real-time interference measurements. Focused real-time interferometric patterns were recorded on a 35-mm photographic film, Fuji Neopan SS-DX (100 mm^{-1}).

To verify the proposed approach and to demonstrate its abilities, a series of diagnostic experiments have been conducted with a shock flow as the phase object. The 400×500 mm^2 supersonic wind tunnel was used to produce the supersonic flow field to be studied. A spherically symmetric sharp-edged model ("flying plate") at a zero angle of attack served to generate the shock flow field variations (see Figure 4.5, where, with a continuous white light source, Schlieren photographs were also recorded under the same conditions in order to demonstrate general characteristics of the flow field).

Typical interferograms are presented in Figure 6.2. The technique visualizes density gradients in the direction perpendicular to the undisturbed velocity vector. The interferograms were observed in one of the coincident zero-first-positive diffraction orders. It was confirmed experimentally that the contrast of the displayed interference picture is smaller than unity.

It was also confirmed that, in comparison with classic and holographic reference beam techniques, the novel method is more tolerant to relatively high mechanical disturbances of the interferometer's scheme during the facility run.

It is believed that the proposed approach will become an useful tool for visualization and accurate mapping of the density gradients of

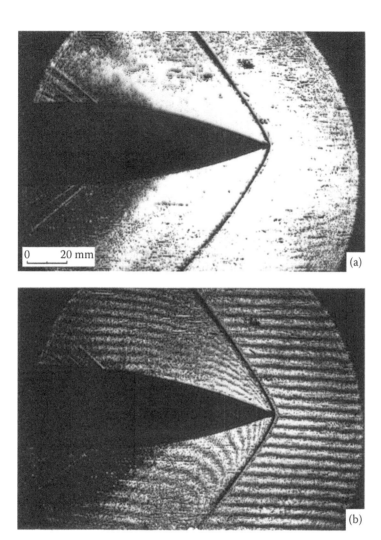

FIGURE 6.2
Focused image real-time shearing interferograms. Interferograms: (a) infinite-width-fringe mode; (b) finite-width-fringe mode. Interferograms were recorded on a Fuji Neopan SS-DX photo film with time exposures ~1 ms by using the light of a He–Ne laser. Mach number $M =$ 1.7. Shears (in the field of vision) are $\Delta x = 0.48$ mm.

gas-dynamic flow fields in wind/shock tunnels, where acoustic noise problems may dramatically affect reference-beam holographic schemes. The technique can also be successfully used in any wide-aperture, powerful facility. Gas-dynamic laboratories where a Schlieren device and a pulsed laser light source are available can, with only a small investment, modify the Schlieren system for the proposed real-time holographic shearing interferometer.

6.2   Disappearance of Holographic and Interference Fringe Accompanying Optical Diagnostics of a Supersonic Bow Shock Flow

Preliminary results on optical diagnostics of bow shocks in a supersonic wind tunnel by applying a dual hologram shearing interferometry technique are discussed in this section. A strong refraction effect of the probing beam penetrating the region in the vicinity of a bow shock over a blunt nose cone model has been discovered.[77] In a signal hologram, the effect leads to the disappearance of holographic fringes in a narrow region attached to the shock wave front. The reconstructed interferogram in this region manifests the absence of an interference pattern.

Computer simulations were performed for a part of the probing beam penetrating the area of high-density steep gradients of the compressed air attached to the central part of the shock front of a bow shock. The compressed area was modeled as a hyperbolic cap. The bow shock was assumed to be axisymmetric. The simulations made it possible to evaluate angles of deflection, and found conformity with reconstructed interferograms (shadowgraphs).

It is concluded that, in the above-indicated region of bow shock flows, the probing light deviates refractively into some angles that could be large enough for light rays to be blocked out and never arrive at the detector (photographic film). In the case when interferometric fringes disappear, the effect of strong refraction makes it impossible to measure air density gradients in some critical regions.

6.2.1   Introduction

Modern experimental studies of supersonic and hypersonic compressible flows require more detailed data than before, mainly for the purposes of validation, verification, and calibration of computational fluid dynamics (CFD) calculations. As to nonintrusive tools developed for subsonic incompressible flows, such as particle image velocimetry (PIV) and laser Doppler anemometry, they are not suitable for high-speed compressible flows. Hot wire, hot film techniques, and Pitot tube surveys introduce strong disturbances, especially when the probes are submerged in a flow field in proximity to areas with discontinuities such as shock flows.

Nonintrusive optical diagnostic techniques have remained one of the main tools for extracting detailed information in this region of compressible flows. Quantitative data can be obtained by Schlieren,[78] classic interferometry,[79] reference beam,[80] and shearing holographic techniques.[57]

It is well known that supersonic and hypersonic wind tunnel facilities during a run effectively generate acoustic disturbances. Unfortunately, reference beam classic and holographic interference schemes demonstrate a low mechanical stability and do not tolerate even a low level of acoustical disturbances. Classic shearing schemes are more stable,[62] but in general possess most of the shortcomings inherent in reference beam systems. Over approximately the last three decades, a dozen acoustically tolerant optical schemes have been suggested. Unfortunately, these schemes possess different shortcomings.

The main disadvantages of such schemes are as follows: (1) inability to record finite-width-fringe interferograms; (2) necessity to use an expensive pulsed laser system for very short exposure times; (3) necessity to use high-quality optics, because the interferograms obtained are not compensated for aberrations; and (4) low acoustic stability because of remarkably spatially separated reference and object beams. The diffraction interference scheme (Figure 4.1) with large-aperture Schlieren mirrors is well suited to various gas-dynamic and high-power facilities. The advantages are the following: (1) simplicity of the optical scheme, (2) a high spatial resolution which is comparable with resolution of Schlieren systems, and (3) a high mechanical stability comparable with the stability of Schlieren systems. Tolerance to strong acoustical disturbances, which always accompany a wind tunnel run, is high enough for recording signal holograms by using a continuous wave laser system. Utilizing a phase grating with relatively low spatial frequencies (2.5, 5.0 mm^{-1}) makes it possible to use a series of photo films with resolving limits ~100 lines/mm.

Experimental applications of the proposed technique are accompanied by some problems, which will be demonstrated below in examples of interference diagnostics of a supersonic bow shock. The main difficulties arise due to effects of refraction in the area attached to the front of the bow shock, which is characterized by a strong degree of compression and gradients. Strong steep air density gradients lead to remarkable refraction of a signal wave in this region and deflection of rays from their straightforward propagations. Some rays cannot arrive at the plane of registration, i.e., at the plane of a photographic film, due to the large arms of the receiver section of the optical scheme and remarkable deflections. Other problems arise due to a high level of flow fluctuations in some regions of a flow field, owing to excessive lateral shears, and may dramatically affect the contrast of interference fringes. Both effects lead to disappearance of the interference pattern. In the first case holographic fringes also disappear, whereas in the second case the contrast of interference fringes dramatically drops.

6.2.2   Reconstructed Interferograms

In Figure 6.3 and Figure 6.4, a series of reconstructed interferograms for Mach numbers $M = 1.6$ and 2.5, supplemented with correspondent shadowgraphs, are presented. A shadowgraph can be easily reconstructed from a signal hologram only. Figure 6.3(a–d) demonstrates the situation for the weakest degree of compression, where only the narrow ~1 mm area attached

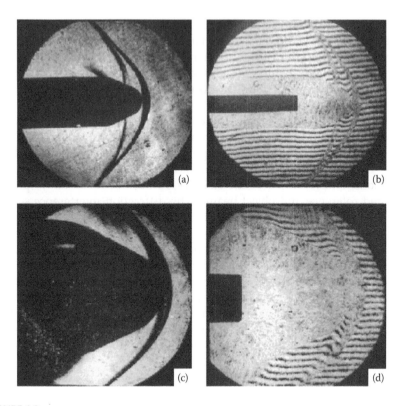

FIGURE 6.3
(a, c) Reconstructed shadowgraphs and (b, d) shearing interferograms at Mach number $M = 1.6$. Shears: 0.2 mm. Coefficient of magnification is (a, b) 0.2× and (c, d) 0.475×.

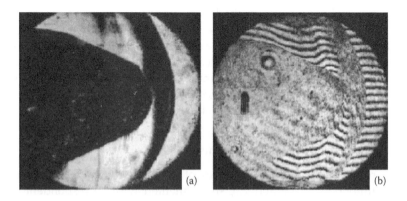

FIGURE 6.4
(a) Reconstructed shadowgraph and (b) shearing interferogram at Mach number $M = 2.5$. Shears: $s = \Delta x = 0.2$ mm. Coefficient of magnification is 0.475×.

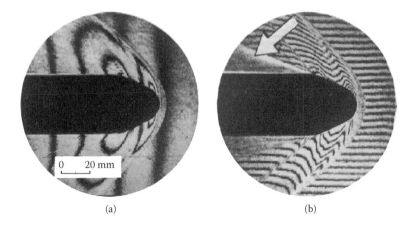

FIGURE 6.5
Reconstructed shearing interferograms at Mach number $M = 2.0$

to the top of the model does not have holographic fringes. Regions with high steep density gradients, and consequently remarkable refraction, are clearly visible in the reconstructed shadow picture (shadowgraph) (see Figure 6.3[a] and [c]). These regions are attached to a bow shock.

In Figure 6.4(a) and (b) bow shocks at $M = 2.5$ can be observed. The regime with $M = 2.5$ maximizing compression downstream of a bow shock is characterized with further spreading of the region with disappearance of interference fringes.

In the region between the top of the model and the central part of a bow shock, holographic fringes are not observed. A flow field is presented in Figure 6.5(a) and (b) with mediate Mach number $M = 2.0$ and shears: (a) $s = \Delta y = 0.1$; (b) $s = \Delta x = 0.2$ mm. The coefficient of magnification is 0.2×. The arrow shows the region with a dramatically dropped contrast of interference fringes. Regions with disappeared interference due to refraction are also observed. Interesting behavior of an interference pattern could be observed in regions with a high level of fluctuations, if the experimentally chosen lateral shear is in excess of some value. Shearing interferometry is more attractive and flexible in comparison with its reference beam counterpart, because it allows changing the sensitivity of interference measurements. The sensitivity could be enhanced due to enlargement of a lateral shear, s. If this value is still small, it is permitted[81] to write the fringe equation in the form:

$$2\pi N\left(x,y\right) = ks\frac{\partial \Phi}{\partial x} = \int_{z_1}^{z_2} \frac{\partial}{\partial x} K\rho\left(x,y,z\right) dz \tag{6.6}$$

where $N(x, y)$ is fringe shift, k is the propagation vector, Φ is phase, K is the Gladston–Dale constant, and ρ is air density. Thus, the fringe shift is directly proportional to the lateral shear, s. In Reference [82], it was found experimentally that the degradation of the contrast occurs due to excessive lateral shear. At the same time, in Figure 6.5(b), this region is absent due to smaller lateral shears ~0.1 mm. Thus, we have discovered two experimentally observed mechanisms of lost interference fringes in some regions of a bow shock, diagnosed by a lateral shear holographic technique. The first mechanism is determined by the effect of a strong refraction for some rays traveling in compressed deep gradient regions. The second mechanism is connected to regions with strong flow fluctuations and excessive lateral shifts.

6.2.3 Microscopy Investigation

Microscopy made it possible to accurately study the structure of a shock flow and measure spatial frequencies of signal holograms. It was found that shearing holograms were recorded with the widths of holographic fringes from ~5.5 to 22.2 μm. Interferograms were usually recorded on a photo film with magnification 0.2×. Special measures were particularly taken for enhancing the spatial resolution of the area under interest. Thus, the lowest spatial frequencies correspond to signal holograms, which were recorded with maximum possible magnification equal to 0.475×. The magnification is limited by the size of the frame of a photographic film 24 × 36 mm². Microscopic observations also allowed for determining and measuring the geometrical sizes of the area attached to the front of a bow shock, which is free of holographic fringes. This area around the aerodynamic model is well suited to the area visualized on reconstructed shadowgraphs. Besides that, in signal holograms a double structure of this area is seen. The area is forwarded with a thin ~0.2 mm layer, which is attached just to the front of a bow shock. This layer corresponds to the distance ($s = \Delta x = 0.2$ mm) between fronts of bow shocks arising due to lateral shear in the horizontal direction. Microscopic testing of a signal hologram in the vicinity of the top of the aerodynamic model shows that the area attached to a bow shock in the central and peripheral regions is free of holographic fringes.

6.2.4 Simulations of Refraction of Diagnostic Rays in a Bow Shock

Ray propagation was calculated in the vicinity of the central part of a bow shock using a numeric solution of the differential equation of light rays:[83]

$$\frac{d}{ds}\left\{ d\frac{d\vec{r}}{ds}r \right\} = \nabla n \tag{6.7}$$

where $r(x, y, z)$ is a position vector of the point on a ray, $n(x, y, z)$ is the refractive index, and s is the length of the ray measured from the fixed point on it. The form of shock wave front with a hyperbolic function was fit:

$$x + a_0 \left(y^2 + z^2 \right)^{1/2} = p + p_0 \qquad (6.8)$$

where a_0 and p_0 are fitting parameters, and p describes the position where the hyperbola crosses the x-axis ($y = z = 0$). It was assumed, for calculation, that the refractive index distribution around the shock wave is a p-dependent function; $n(p)$ is shown in Figure 6.6. The total angles of beam deflection are also shown in Figure 6.6. It is seen that maximal deflection angles are in dozens milliradian range. These angles are approximately one order of magnitude bigger than characteristic angles of diffraction; thus the diffraction can play some role just in front of the model where the deflection angles change substantially, and the body is in close proximity. Figure 6.6 confirms in general outline the characteristics of the reconstructed interferograms shown in Figure 6.5.

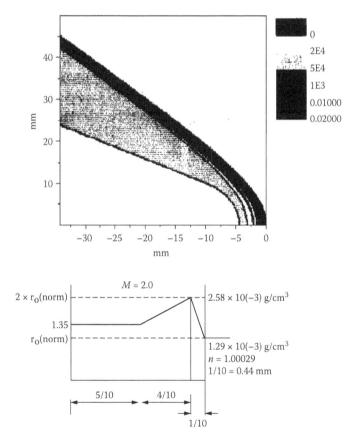

FIGURE 6.6
Calculated full angles of diagnostic rays, deflected from their straightforward propagations at $M = 2.0$ (above); one of simplified reasonable distributions of the refractive index (below).

6.2.5   Discussion and Interpretation of Obtained Experimental Results

Optical diagnostics by methods of holographic shearing interferometry have some principal advantages in comparison with classic shear and reference beam holographic techniques. The method is useful for studying some complicated aerodynamic objects that significantly refract a diagnostic beam, for instance, supersonic shock flows over a blunt body accompanied by effects of disappearing interference fringes. Actually, disappearing interference fringes lead to the situation that makes it impossible to calculate a density distribution in these regions. Disappearing interference fringes in regions with a high level of flow fluctuations might be overcome by minimizing possible lateral shear, as is seen in Figure 6.5(a). At the same time, holographic fringes are present at the interferogram imaged in Figure 6.5(b) with some diminishing of the contrast. It is interesting to note that the effect of dramatically changing the contrast of interference patterns might be used to reveal such regions simply by enlarging a lateral shear. Disappearance of interference fringes (and primarily, holographic fringes) due to effects of refractions is more difficult to overcome. In fact, the two-pinhole diaphragm (16) in Figure 6.5 works as a spatial filter for the rays encountered at some insignificant angles of refraction (i.e., only these rays are selected). As to the rays refracted at relatively large angles, they are blocked out. Thus, the scheme of a holographic interferometer is only well suited for phase objects.

To accurately diagnose aerodynamic transparent objects, some construction changes must be introduced into the scheme. As the first step, a phase grating with a larger spatial frequency should be chosen in order to increase the distance between diffraction orders in the plane of a two-pinhole diaphragm. The second step is to enlarge the diameters of the pinholes to select not only practically undisturbed rays, but also the rays affected by some considerable refraction. On the other hand, a two-pinhole diaphragm could be used for the accurate separation of rays refracted at different angles visualizing the picture for different inclinations. In the literature, the problem of measuring ray inclinations was discussed in Reference [84], where deflections were measured by the speckle technique. The lowest boundary between phase and transparent objects was taken to be of the order of 0.1 mrad.[85] On the other hand, this value could be a few times larger (~50 mrad).[86] Taking into account the algorithm of properly changing the design of a receiver section, the holographic shearing interferometer could be recommended for various aerospace engineering applications due to a simple but mechanically tolerant optical scheme, which could be designed with a small investment on the base on an ordinary Schlieren system.

Conclusion to Section II

Mach–Zehnder shearing holographic interferometry was successfully applied for studying shock flows over different aerodynamic models in a supersonic wind tunnel. The proposed optical scheme of a diffraction shear interferometer demonstrates a high tolerance to acoustical disturbances of powerful aerodynamic facilities. A high level of stability allows recording of signal holograms during the run of a facility using CW helium–neon and argon ion lasers. Mach–Zehnder shearing holography demonstrates all advantages characteristic of reference beam holographic interferometry, including the opportunity to improve the sensitivity of interference measurements.

The optical scheme of the diffraction holographic interferometer could be constructed on the basis of a Schlieren device; at the same time aberrations in output reconstructed shearing interferograms are cancelled. The scheme is applicable for different types of powerful strongly noising facilities having a large aperture of test sections: 1×1 m^2 and more. The analysis of reconstructed interferograms shows that the spatial resolution of the technique is partly restricted due to spherical aberrations characteristic of Schlieren mirrors.

The two-wavelength approach makes possible an essential reduction of the the sensitivity of athe interference shearing technique, and thus allows successful measurements of the shock flow over a bowed nose aerodynamic model. The technique is useful in the case of the strong discontinuity for measuring fringe shifts smaller than unity.

It was shown in Chapter 6 that the real-time mode of the interference measurements is applicable to studying a shock flow in aerodynamic facilities. Different phase objects could be investigated in a regime of "living fringe" by using a CW argon ion or helium–neon laser.

Application of one of the related diagnostic techniques—Moiré deflectometry—could help in preliminary investigation of flow fields in powerful wide-aperture aerodynamic facilities, although the spatial resolution of the technique is lower than in the case of holographic shearing interferometry.

The technique allows producing interference measurements in a whole flow field, including zones of shock flows, and makes it possible to measure the gradients of the refractive index of a phase object, together with the calculation of the contrast of reconstructed interference pattern, which is of special importance in the case of studying turbulent objects.

Section III

Mach–Zehnder Digital Holographic Interferometry and Related Techniques

Recording Mach–Zehnder Digital Interferograms/ Holograms on CCD/CMOS Sensors and Their Applications

Introduction to Section III

In Sections I and II the technology and methods of recording optical Mach–Zehnder reference beam and shearing holograms on photographic films have been accurately analyzed for the purpose of using these two optical approaches for diagnostics of phase objects in the fields of noncoherent laser–matter interaction and gas-dynamic applications. It was shown that the recording holograms and subsequent optical reconstruction and studying signal waves are not cumbersome and are effective for retrieving phase data on different phase objects.

At the same time, the epoch of photographic films is coming to an end, and photographic materials are gradually disappearing from experimental laboratories. In the modern investigations of phase objects, photographic films are every day more and more actively replaced by electronic charge coupled device (CCD) and complementary metal oxide semiconductor (CMOS) sensors. Indeed, Mach–Zehnder holograms may be recorded on the sensitive areas of CCD/CMOS sensors, omitting the photochemical wet processing. From a theoretical point of view, digital holograms/interferograms (CCD/CMOS holograms) can be also related to Mach–Zehnder type holograms because their carrier spatial frequencies do not exceed a few dozens of mm^{-1}. As a rule, resolving powers of most sensors correlate with this value.

Recording Mach–Zehnder holograms on CCD/CMOS sensors, so-called digital holographic interferometry, possesses remarkable advantages over optical recording on photographic films. Besides omitting photochemical wet processing, preliminary recordings of signal and comparison holograms on CCD/CMOS cameras could be transferred in digital form to the hard drive of a computer. Thereafter, the holograms may be digitally processed for the purpose of digital reconstruction of interferograms. The digitally reconstructed interferogram facilitates retrieval of interference data and calculation of the refractive index of the phase object under test (or its gradients), including the contrast of the interference pattern.

The approach is not cumbersome and time consuming; imaging a phase object on the sensitive sensor of a CCD/CMOS camera is a simple and quick operation. It is almost ideally suited for an express analysis of the phase object under test. From numerically reconstructed interferograms, the experimenter can get the amplitude and phase data very quickly, because all steps are computerized.

The quick progress in the creation of full-frame (wide-aperture) sensors will supply the researcher with the technique of studying different phase objects with a low level of aberrations and a high spatial resolution, especially in the case of studying aerodynamic flow fields in wide-aperture shock/wind tunnels. In Section II it was shown that instability in the flow field under study or its turbulence leads to the reduction of the contrast of reconstructed (recorded) interferograms. From this point of view, digital holographic interferometry could be the ideal tool for determination of the level of a turbulence/instability. This approach is also effective in the case of studying turbulent plasmas.

It should be emphasized that digitally reconstructed interferograms are aberration free, owing to the differential character of holographic interferometry; i.e., aberrations caused by imperfections of the holographic interferometer may be cancelled digitally, as was demonstrated in the case of optical holograms recorded on photographic films in Chapter 2. As to digitally recorded reference beam and shearing interferograms, their aberrations are not compensated.

Classic interferograms could be successfully digitally processed if the useful fringe shifts caused by phase objects noticeably exceed the level of the visibly unresolved fringe shift. It should be noted that classic digital interferograms and holograms are practically indistinguishable from the point of view of carrier spatial frequencies, unlike the case of optical Mach–Zehnder interferograms/holograms recorded in finite-width fringes on photographic films.

The methods of numerical processing of interferograms reconstructed from digitally recorded reference beam holograms are carefully described in two fundamental books.[87,88] The interested reader will find in the books the detailed procedures of retrieving interference data for the purpose of obtaining 3-D distributions of the refractive index. As to digital recording and computer processing in the case of the shearing approach, when the first derivative of the refractive index is calculated, the corresponding information might be found in References [89, 90]. Taking into account these speculations, the author will pay little attention to the numerical processing and retriee of interferometric data. Attention will be paid to special features of the procedure of imaging of phase objects on the CCD/CMOS sensors of cameras. In Chapter 7, holographic interferometers for digital reference beam and shearing Mach–Zehnder holograms will be analyzed separately, as in the case of optical Mach–Zehnder holographic interferometry. Examples of digital imaging of different phase objects under test on CCD/CMOS cameras will be demonstrated.

In addition to digital interferometric methods of investigation of the phase, there are other related diagnostic techniques in experimental practice. The main related techniques, which will be analyzed in Chapter 8, are Schlieren graph and Moiré deflectometry. Both methods measure gradients of the refraction index: (1) the Schlieren technique in the direction perpendicular to an optical knife; and (2) the Moiré deflectometry in the direction perpendicular to the Moiré fringes.

The Schlieren technique is characterized by a simple optical scheme, a high mechanical stability, and a relatively high spatial resolution depending, in the case of using photographic films or CCD/CMOS sensors, on the resolving power of a focusing objective lens. As to the Moiré technique, it in some respects is similar to Mach–Zehnder shearing holographic interferometry. The difference is in lower spatial resolution in comparison with the holographic technique.

In Chapter 9, principles of spatial filtering of a narrow pencil beam of continuous wave (CW) lasers, special features of spatial filtration applicable for holographic interferometers, and schemes of reconstruction of Mach–Zehnder holograms will be also discussed.

Finally, devices and equipment required for performing routine Mach–Zehnder holographic interference experiments will be analyzed.

7

Mach–Zehnder Digital Holographic Interferometry

7.1 Main Technical Characteristics of CCD/ CMOS Sensors Responsible for Imaging Digital Mach–Zehnder Holograms

7.1.1 Algorithms of Digital Recording and Important Technical Characteristics of CCD/CMOS Sensors

The technique of imaging the interference pattern of a phase object on the sensitive area of a CCD/CMOS sensor has numerous substantial advantages over imaging the holograms/interferograms on photographic films. The advantages are obvious: the acquisition of the digital phase information on the phase object by means of recording a solitary interferogram or a series of consequent interference images and saving them on the memory card of a CCD/CMOS camera. A memory card or flash card is an electronic flash memory data storage device used for storing digital information. The digital information recorded on a memory card may be subsequently transferred to the hard drive of a computer. The digital holograms/interferograms can be numerically processed on a computer by using, for example, MATLAB® or other software applicable for the numerical processing of rectangular pixel matrices. Digital processing allows for the reduction of interferograms/holograms for the purpose of numerical retrieval 3-D distribution of the refractive index and /or the contrast of the phase objects under test.

At the same time, the digital approach makes it possible to display the focused image Mach–Zehnder holograms/interferograms on the monitor of a computer for visualizing the phase object under study in pixel form. The imaging of focused interferograms/holograms is realized by means of using high-quality CCD/CMOS objective lenses.

Limitations arise from the fact that the sensitive areas of conventional CCD/CMOS sensors have pretty small apertures. The exceptions are CCD/ CMOS sensors with an extra large aperture of the sensitive area, which is comparable with C-format 35-mm photographic film. Examples of some

CCD/CMOS include the 25 × 25 mm² sensor of Tektronics/CCD-1024Tkx or Alpha-900 Sony CMOS camera with the so-called "full frame" sensor with a size of 35.9 × 24 mm². A large aperture is preferable for the purpose of imaging, especially in the case of flow field studies in wind/shock tunnels and other large-aperture facilities, where the imaging is performed with smaller coefficients of reduction ($M \geq 0.1$) and consequently with imaging aberrations which are not substantial. Unfortunately, wide-aperture sensors as a rule have relatively low temporal characteristics.

The other important limitation is the relatively low resolving powers of CCD/CMOS sensors. Generally speaking, imaging Mach–Zehnder interferograms/holograms on CCD/CMOS sensors has the same difficulties as in the case of imaging them on photographic films. CCD/CMOS sensors have a little bit worse resolving powers than ordinary photographic films: ≈50 mm⁻¹. That is why the experimenter should permanently control carrier spatial frequencies of the recorded interferograms/holograms—they should not exceed the resolving power of a CCD/CMOS sensor.

Carrier spatial frequencies of the interferometric/holographic patterns are of the same order. A recorded interferometric pattern in finite-width-fringe mode is called a "hologram" (signal hologram), if a comparison hologram is also imaged together with it on the sensor. The signal hologram is digitally compared ("mathematically subtracted") with the comparison one for the purpose of numerical calculation of the reconstructed interferogram. Note that the numerically reconstructed digital interferogram is free of aberrations of the recording setup owing to the differential character of holographic recording (see Chapter 2). In the case of recording classic interferograms, aberrations of the recording setup are not compensated; therefore the useful fringe shift should be remarkably larger than the visually unresolved fringe shift, $\Delta N_v = 0.1$, in order to perform the interference measurements with a high accuracy.

As already stated, *spatial resolving powers* of most CCD/CMOS sensors, are a bit lower or compatible with the spatial resolution of standard photographic films. The spatial resolving power of a CCD/CMOS sensor in terms of spatial frequencies is approximately determined as the spatial frequency of an interference picture at which the modulation transfer function (MTF) of the sensor drops two times: $MTF \approx 0.5$. In this case, the recorded spatial frequencies of holographic pictures for most CCD/CMOS sensors should be smaller than ≈0.25 lines × pix⁻¹ (four pixels per fringe).

There are sensors having extraordinary resolving powers. For example, the Sony CMOS sensor IMX017CQE has a pixel size of 2.5 μm. With this pixel spacing of 2.5 μm, we could expect to resolve fringes with the spacing 4 × 2.5 μm = 10 μm. Probably, we could expect to begin resolving black/white lines with one half of that spacing: 5 μm is the equivalent linear resolution. This value corresponds to ~100 mm⁻¹ for the resolving power of the sensor.

CCD/CMOS imaging is usually produced by the sensors having *the size of a pixel* of the order of ~5 to 10 microns; thus the linear spatial resolution of a sensor can be evaluated for ~0.25 fringes per pixel as 10 to 20 µm. The resolving powers are ~50 to 25 mm^{-1}.

The size of an imaging matrix is characterized by the factor of the *pixel resolution*. The standard values, for example, are 640 × 480 for typical VGA format sensors. Megapixel cameras may have pixel resolution 1680 × 1120 and higher. Aspect ratios are 16:9 or 3:2. The Sony CMOS sensor IMX017CQE has a pixel resolution of 2921H × 2184V.

One of the substantial characteristics of CCD/CMOS sensors is their ability to image a high-quality high-*grayscale (shades of gray) level image* on the monitor. Some commonly encountered gray level values are: 256, 1024, and 4096. The number of distinct gray levels is usually a power of 2; that is, the gray level = 2^B, where B is the number of bits in the binary representation of the brightness levels.

As to *contrast characteristics* of most CCD/CMOS sensors, they have comparable or even higher contrast than photographic films. They are commercially produced with a large variety of apertures, speeds, and numbers of pixels (resolution), including megapixel systems and other electronic and optical characteristics.

One of their important optical parameter is a wide *spectral sensitivity* from 400 to 1000 nm of standard CCD/CMOS sensors. Standard photo materials cannot guarantee such a wide working spectral range. Orthochromatic materials working in the range of 400 to 550 nm have very low sensitivities in the red visible range and are not sensitive in the near-IR region; as to panchromatic photo materials with spectral sensitivity limits of 400 to 690 nm, they are also not sensitive in the near-IR region.

Temporal resolution of CCD/CMOS sensors is also of high importance for transient phase objects, mainly in the case of the pulsed-periodic working range of a diagnostic laser. On the other hand, low-temporal-resolution CCD/CMOS cameras having, as a rule, high pixel resolutions could demonstrate a high temporal resolution by working in a waiting regime with the single pulse of a Q-switched pulsed laser system.

There are two popular categories of *CCD camera signal standards*: the RS-170, which is the American standard and provides an analog output of 30 fps (frames per second) with each frame, has two interlaced fields; and CCIR, which is the European standard of camera signals with 50 fps. The electronics of conventional cameras provide clock rates of tens of megahertz, allowing reading tens of millions of pixels per second. Usually 30 fps, which means the exposure time is $^1/_{30}$ of a second, is perfectly fit to low-speed phenomena. High-speed transient phenomena can be studied by a fast CCD camera, which is capable of operating at, for example, 40,500 Hz with a controllable shutter speed (Kodak EktaPro fast visible digital video camera). In the case of full resolution (256 × 256 pixels), the camera is capable of acquiring 4,500 frames per second. Faster frame rates provide lower resolution, where a minimum exposure of 20 ns is achievable.

The CPL series of high-speed CCD and CMOS cameras are capable of recording speeds from 1 to 50,000 fps in continuous mode and over 400,000 fps in burst mode. Megapixel CMOS sensors, with 1280 × 1024 pixels, can work at speeds of 50,000 frames per second over a wide spectral sensitivity range, 400 to 1000 nm, in an 8-bit dynamic regime. The electronic shutter speed varies from 0.5 μs to 30 ms in 1-μs steps. Such cameras allow the imaging movie-interferometry of numerous transient phase objects in aerodynamic research, such as turbulent flows, droplet atomization, different types of flames, subsonic flow fields, etc.

In conclusion, it should be noted that, because of a rapid advance in CCD/CMOS technology, there would be a lot of interest in recording the whole sequence of interferograms/holograms with a Mach–Zehnder configuration in order to obtain the movie interferometry of the transient phase objects under investigation. This approach is determined first of all by the temporal characteristics of CCD/CMOS systems. In contrast to imaging interferograms on photo materials, optical setups of the interferometer plus imaging CCD/CMOS cameras have better mechanical stability because of the use of noiseless electronic shutters. Thus, CCD/CMOS imaging has numerous advantages over imaging on photographic films.

The author has successfully applied a high-contrast VGA-CCD sensor with a resolution of 640 × 480 pixels and an aperture of 6.4 × 4.8 mm^2 for wind-tunnel testing. A similar type of CCD system was used for interference applications, namely a HTC-550 series CCD camera with an image area of 7.95 × 6.45 mm^2; the resolution of the camera is 811(H) by 508(V). A similar CCD-sensor with 512 × 512 pixels has been used for interference measurements on turbulently vaporizing droplets. Examples of applications of digital imaging of the phase objects under test will be demonstrated below.

7.1.2 Ensemble of CCD/CMOS Camera and Its Objective Lens

In the problem of digital focusing, the location of the focusing CCD/CMOS-lens plays an important role in the scheme of recording Mach–Zehnder reference beam interferograms/holograms. CCD/CMOS cameras are supplied with high quality objective lenses, which facilitate recording fine phase details of an object on the sensor. Unlike optical holographic imaging on standard photographic films, where the resolution of fine details of the phase object under test depends on the carrier spatial frequency of the signal holograms, in the digital imaging the spatial resolution depends on the ensemble of a CCD/CMOS sensor plus an imaging objective lens. The resolving power of the ensemble could be found in accordance with Equation (2.2). Usually objective lenses of CCD/CMOS cameras have better resolution characteristics in comparison with sensors; thus the resolution of an image is defined mainly by the spatial characteristics of the sensor of a CCD/CMOS camera. The exception is the Sony sensor IMX017CQE which has a resolving power ~100 mm^{-1}. It is hard to find a CCD/CMOS lens with better resolving power.

As a rule, CCD/CMOS objective lenses are intended for imaging an object at relatively large working distances (distances between the objects under test and the objective lens). Using CCD/CMOS cameras together with the setup of Mach–Zehnder interferometer requires, in the majority of cases, larger optical magnifications of the object. From an optical point of view, this means the minimization of the distance between the imaging objective lens (the first objective lens of the collimator [see Figure 1.1]) and a phase object. This circumstance is in contradiction with the long working distance characteristic of objective lenses attached to CCD/CMOS cameras. The discrepancy is removed by disconnecting the objective lens attached to a camera and installing it at some reasonable distance from the camera. The disconnected objective lens could be attached to a threaded or bayonet ring mechanical holder. The other approach is to use the lens as the second objective lens of the collimator.

The third option is connection of an adaptor between these two optical elements. The adaptor increases the distance between the CCD/CMOS objective lens and the camera. CCD/CMOS lenses are usually available in two different lens mounts. "C-mount" lenses have a flange back distance of 17.5 mm, while a "CS-mount" lens has a flange back distance of 12.5 mm. The flange back distance is the distance from the flange of the lens (beginning of the lens mount) to the focal plane. In other words the distance, 17.5 mm (or 12.5 mm), is determined as the distance from the sensor surface to the outer face of the thread barrel. The adapter makes it possible to image phase objects that are located at shorter working distances than the standard working distance of CCD/CMOS lenses.

7.2 Using CCD/CMOS Sensors for Recording Digital Mach–Zehnder Reference Beam Holograms/Interferograms

Careful description of principles and methods of digital recording and computer processing of interferograms/holograms is presented in the fundamental books.[87,88] The basic algorithms of digital recording and processing were finally formulated in References [89–91].

A new method of phase determination in hologram interferometry is discussed in Reference [89]. The Fresnel holograms, which represent the undeformed and the deformed states of the object, are generated on a CCD target and stored electronically. No lens or other imaging device is used. The reconstruction is done from the digitally stored holograms with mathematical methods. It is shown that the intensity as well as the phase can be calculated from the digitally sampled holograms. A comparison of the phases of the undeformed and the deformed states permits direct determination of the interference phase.

Reference [90] describes the principles and major applications of digital recording and numerical reconstruction of holograms (digital holography). Digital holography became feasible after charge coupled devices (CCDs) with suitable numbers and sizes of pixels, as well as computers with sufficient speed, became available. The Fresnel or Fourier holograms are recorded directly by the CCD and stored digitally. No film material involving wet-chemical or other processing is necessary. The reconstruction of the wave field, which is done optically by illumination of a hologram, is performed by numerical methods. The numerical reconstruction process is based on the Fresnel–Kirchhoff integral, which describes the diffraction of the reconstructing wave at the microstructure of the hologram. In the numerical reconstruction process, not only the intensity, but also the phase distribution of the stored wave field can be computed from the digital hologram. This offers new possibilities for a variety of applications. Digital holography is applied to measure shape and surface deformation of opaque bodies and refractive index fields within transparent media. Further applications are imaging and microscopy, where it is advantageous to refocus the area under investigation by numerical methods.

A coherent optical imaging system consisting of recording a digital hologram by a CCD array and numerical reconstruction of the complex wave field in a computer is subjected to a frequency analysis in Reference [91]. This analysis recognizes Fresnel and lensless Fourier transform holography, collimated and divergent reference waves, as well as the real image, the virtual image, and the dc term. The influences of finite sampling and the fill factor of the CCD pixels are examined. The impulse response of the system is a shifted Fraunhofer diffraction pattern of the aperture defined by the CCD. A fill factor below unity leads to a contrast decrease, which is quantitatively described in the modulation transfer function.

In the present section the author first of all would like to direct the attention of the reader to engineering problems of digital recording classic interferograms and reference beam holograms. The Mach–Zehnder optical scheme is presented and analyzed as the basic interferometric system for recording the phase information on a phase object. Section I contained a preliminary discussion of imaging and recording optical interferograms and holograms on photographic films. Let us consider the digital recording on electronic CCD/CMOS sensors.

7.2.1 Using Reference Beam Mach–Zehnder Interferometer for Recording Digital CCD/CMOS Holograms

It is interesting to note that the optical setup imaged in Figure 1.1 is also well suited for recording digital Mach–Zehnder interferograms/holograms on the sensitive areas of CCD/CMOS sensors. In that case, the photographic film (16) is replaced by a CCD/CMOS sensor. The disconnected original objective lens of a CCD/CMOS camera can be included in the collimator (13, 14). The

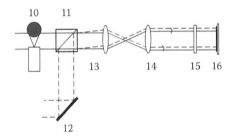

FIGURE 7.1
Receiver of Mach–Zehnder interferometer with recording on the CCD/CMOS sensor of a CCD/CMOS camera. (10) Phase object under test; (11) beam-splitting cube in object arm; (12) flat mirror in reference arm; (13), (14) collimator; (15) filter; (16) CCD/CMOS sensor.

receiver of the Mach–Zehnder interferometer scheme is imaged separately in Figure 7.1 in order to clarify the approach. As was noted in Section 7.1, the distance between the phase object (10) and the first objective lens (13) should be minimized to realize the maximum possible coefficient of magnification. In practice, this distance is restricted by the size of an optical table and the beam-splitting cube (11) on it. The next interconnected imaging problem is shortening the working distance of the objective CCD/CMOS lens that was discussed in Section 7.1.

The principal positive step for the experimenter is as follows: he/she should strive to work with the camera having the largest possible aperture of the sensor, because the larger optical reduction leads to larger optical aberrations, which distort the image of a phase object and deteriorate the spatial resolution of the image. If the aperture of the beam-splitting cube (11) is one inch or larger, then the best choice is using a full-frame camera having the sizes of sensors close to the C-frame of a standard 35-mm photographic film, which is 36×24 mm^2 (Sony α DSLR-A900 [2008], Canon EOS 5D Mark II [2008], Nikon D3X [2008], Sony α DSLR-A850 [2009], Nikon D3S [2009]).

For example, using the Sony Alpha-900 camera having CMOS sensors with an aperture of 24×35.9 mm^2 (demonstrating very high resolution up to 25 megapixels), the experimenter may image a phase object in a 24-mm-diameter spot on the sensor. The sensor is characterized by a resolving power ~42 lines/mm (MTF ~0.5), which means that if a holographic/microinterferometry Mach–Zehnder picture has a spatial frequency less than 42 mm^{-1}, it may be successfully recorded on the sensor. The size of the pixel is 5.925 microns, which means an ~12 μm linear spatial resolution (magnification is unity). In any case, the original CMOS objective lens possesses better spatial characteristics; therefore the spatial resolution of the focused imaged phase object depends on the spatial resolving power of the sensor.

Unfortunately large-aperture sensors have low temporal characteristics. In burst regime, the Alpha-900 camera takes pictures at the speed of 5 frames per second. This is the maximum possible speed for making pictures in the regime of cine (movie) interferometry/holography. In a waiting regime with an the open shutter, the camera may be used for measurements with nanosecond temporal resolution by applying a Q-switched pulsed laser system working in the visible range.

Historically, the process of studying digital recording and digital processing has started with recording the phase objects on photographic films and mathematical processing of recorded interference patterns.

7.2.2   Recording Interferograms/Holograms on Photographic Films and Their Numerical Processing

The first articles on imaging reference beam interferograms on photographic films followed by numerical processing on a computer were written in the 1980s.[92–97] The fruitful approach to retrieving the interference data recorded on a photographic film was based on ideas of a fundamental article,[92] where for the first time the analysis of the Fourier transform method by numerical processing of a finite-width-fringe interferogram was proposed. The fast-Fourier-transform method of topography and interferometry was realized by taking interferometric pictures on a photographic film. By computer processing of a non-contour type of fringe pattern (finite-width-fringe pattern), automatic discrimination is achieved between elevation and depression of the object or wave front form, which was not possible using the fringe-contour-generation techniques. The method has advantages over Moiré topography and conventional fringe-contour interferometry in both accuracy and sensitivity. Unlike fringe-scanning techniques, the method is easy to apply because it uses no moving components.

In Reference [93], two-dimensional sinusoid fitting and Fourier transform methods of analyzing fringes to determine the wave front topography are described. The methods are easy to apply because they do not require finding fringe centers and fringe orders. Also, they are accurate. For an active optics experiment in which the authors used these techniques, experimental noise exceeds the error resulting from analysis of noise-free theoretical fringe patterns.

A method of analyzing interference fringes using a microcomputer is proposed in Reference [94]. Interference fringes with a tilted wave front are spatially scanned by a TV camera perpendicular to the spatial carrier. Irradiance distribution along a raster scan is sinusoidally interpolated in each carrier interval, and optical phases are calculated over the whole area under test.

In Reference [95], the Fourier transform based on the interferogram analysis technique of Takeda et al.[92] is reviewed, and it is shown that the digitization of the interferogram can act to seriously distort the recovered phase object. An algorithm for dealing with this problem is proposed. The recovered phase object may also be seriously distorted if the response of the recording medium is nonlinear. An algorithm for calibrating the film response is proposed which uses only the data in the interferogram, and which can be used to correct the effects of the film nonlinearity. The technique is demonstrated on experimental interferograms obtained in a study of laser-produced plasmas.

A method for accurate phase determination in holographic interferometry using a one- or two-dimensional Fourier transform is described in Reference [96]. The method calculates the interference phase pointwise, even between fringe extremes, and thus has advantages over conventional fringe-finding and fringe-tracking methods. Only one interference pattern may be used, although the use of two patterns reconstructed with a mutual phase shift permits an easier phase unwrapping and determination of non-monotonic fringe-order variations. Additionally, the method offers a means for filtering out disturbances such as speckle noise and background variations.

The tolerance of the Fourier-transform method of fringe pattern analysis to increasing levels of signal,[97] independent random additive noise, and increasingly complicated phase functions is investigated. It is shown that the condition that the phase must be a slowly varying function compared to the function introduced by the carrier frequency is a fairly flexible one.

Interested readers can find specific information regarding tomographic investigation of phase objects in Reference [98], where two concepts of digital holographic tomography (DHT) based on interferometric tomography and digital holography are presented. DHT enables investigation of phase and amplitude-phase objects. The capability of registration of an object in several views during one registration is the basic advantage of DHT. This is performed by a multiple pass arrangement. Initial experimental results for simple amplitude and phase objects are presented.

It should be remarked that reference beam digital holographic interferometry may be applied to the phase objects having sizes not exceeding the size of the beam-splitting cube. Large-aperture flow fields and plasma objects in wind and shock tunnels and powerful plasma facilities could be investigated by applying the digital holographic shearing interferometry approach. This will be carefully analyzed in Section 7.3.

7.3 Using CCD/CMOS Sensors for Recording Mach–Zehnder Shearing Holograms/Interferograms

The classic shearing interferometry technique has been successfully applied (for overview see Reference [99]) for testing optical components,[100–102] flow investigations,[103] plasma physics,[104] and heat transfer problems.[105]

Specific features characteristic of Mach–Zehnder shearing digital interferometry and digital methods for the investigation of phase objects are discussed in Reference [106]. The shearing setup used a Mach–Zehnder interferometer with objects that were located outside. For the case of radially symmetric phase objects, a numerical method is presented for calculating the gradients of refractive index distribution from data measured by the shearing interferometer. In this type of shearing interferometer, the diagnostic wave is divided into two identical beams behind an object, which are then put on the camera with a small displacement. As was noted, one of main advantages of shearing interferometry is the variable sensitivity, which can be adjusted by changing the displacement between two interfering beams. The evaluation of the interferograms is done by means of Fourier analysis for calculating the first derivatives of the integral phase shift caused by the object under test.

It should be emphasized that the digital interferometric imaging technique cannot perform full compensations of the aberrations of the scheme of a shearing interferometer. Note that the aberrations are the same level as in the case of the reference beam technique. Only the holographic recording on CCD/CMOS sensors of interferometric patterns of a signal and comparison waves allows for computation of the digital reconstruction interferogram where aberrations are compensated.

The holographic shearing approach is demonstrated in an outstanding article.[107] The fundamentals of digital recording and mathematical reconstruction of Fresnel holograms are described. The object is recorded in two different states, and the holograms are stored electronically with a charge coupled device detector. In the process of reconstructing the digitally sampled holograms, different coherent optical methods such as hologram interferometry and shearography are applied. If the holograms are superimposed and reconstructed jointly, this results in a holographic interferogram. If a shearing is introduced in the reconstruction process, a shearogram appears. This means that the evaluation technique (e.g., hologram interferometry or shearography) can be influenced by numerical methods.

An analogous holographic approach with the compensation for aberration was applied in Reference [108]. Combining the concept of lateral shear interferometry (LSI) within a digital holography microscope, the authors demonstrated that it is possible to obtain quantitative optical phase measurement in microscopy by a new single-image-processing procedure. Numerical lateral shear of the reconstructed wave front in the image plane makes it possible to retrieve the derivative of the wave front and remove the defocus aberration term introduced by the microscope objective. The method is tested to investigate a silicon structure and a mouse cell line.

In Reference [109], digital holographic microscopy enables a quantitative phase contrast metrology that is suitable for the investigation of reflective surfaces as well as for the marker-free analysis of living cells. The digital holographic feature of (subsequent) numerical focus adjustment makes possible applications for multi-focus imaging. An overview of digital holographic microscopy methods is described. Applications of digital holographic microscopy are demonstrated by results obtained from living cells and engineered surfaces.

Based on the concept of digital lateral shearing interferometry (LSI) and the idea of automatic aberration compensation (AAC), a method[110] to eliminate the linear term of the sheared phase map in digital holographic (DH) reconstruction is proposed. The procedures for reconstructing the phase image using the LSI-DH method have been analyzed in detail. The analysis indicates that one-dimensional phase unwrapping must be applied to the original phase map before lateral shearing. Moreover, the zeros pad must be applied to the hologram if the phase difference between the two adjacent pixels of the shearogram map is more than 2π. The computer simulations, which are based on the Fresnel digital holography with premagnification, demonstrate the validity of the proposed method.

Generally speaking, digital shearing interferometry/holography could be realized by using three different approaches. The first one is direct recording of a classic shearing interferogram on a CCD/CMOS sensor,[106] and subsequent computer processing of the interferogram for the purpose of retrieving the phase data on an object. The second approach is the following: an object is recorded in two different states, and the holograms are stored electronically on the memory card of a CCD/CMOS camera; subsequently the signal and comparison (or the two signal) holograms are processed digitally on a computer. If the holograms are superimposed and reconstructed jointly, then the reference beam reconstructed interferogram serves as the source of the phase information on the object. If a shearing is introduced in the reconstruction process, a shearing reconstructed interferogram appears.[107] The third approach is recording the signal and comparison shearing holograms the same way as was proposed for optical holographic shearing interferometry in Reference [82]. Note that in the first approach, aberrations are not compensated; as to the second and third techniques, aberrations are compensated due to the differential character of holographic shearing interferometry.

It should be emphasized that for powerful wide-aperture aerodynamic facilities, the optical setups described above in the first and second approaches of shearing interferometry are not applicable. The optimal version in that case is the diffraction holographic shearing interferometer similar to the optical scheme presented in Section 4.2. As was already reported, in aerodynamic (and other wide-aperture powerful facilities) wind and shock tunnel testing flow field experiments, for studying different types of compressible flow fields, it is more flexible and convenient to use CCD/CMOS sensors having maximum possible apertures of the sensitive areas. Full-frame CCD/CMOS cameras provide wide opportunities for interferometric studies of aerodynamic flows. In the optical scheme in Figure 4.1, the photographic camera (18) for the purpose of digital recording and computer processing aerodynamic phase objects should be replaced by a CCD/CMOS camera.

In Figure 7.2 the author proposes the remarkably simplified diffraction scheme of a shearing holographic interferometer in the case of using a CCD/

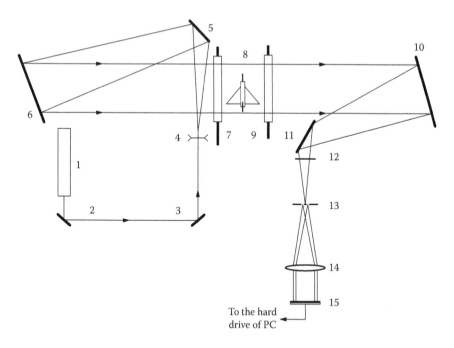

FIGURE 7.2
Simplified scheme of diffraction shearing Mach–Zehnder holographic interferometer for wind tunnel testing. (1) Pulsed laser; (2, 3, 5, 11) flat mirrors; (4) negative lens; (6, 10) spherical Schlieren mirrors; (7, 9) viewing windows; (8) aerodynamic model; (12) phase grating; (13) two-pinhole diaphragm; (14) collimating objective lens; (15) CCD/CMOS sensor.

CMOS sensor. The novel simplified scheme could be designed at the facility of a compressible flow/shock flow based on an approachable Schlieren device. The picture shows the regime with a pulsed laser system as the source of coherent radiation (see Figure 4.1 where a continuous wave [CW] laser is used); the spatial filter is removed and replaced by a negative lens (4). The grating (12) is located in the convergent beam reflected from the second Schlieren mirror (10). A long focal length of the Schlieren mirror (10) guarantees a sufficiently large distance between corresponding circles of confusion on the stop diaphragm (13) in spite of a low spatial frequency of the grating (~2.5 to 10 lines/mm).

Such low frequencies can be explained by the restricted spatial resolving powers of CCD/CMOS sensors. For example, CCD/CMOS sensors of full-frame cameras guarantee a high quality of the interferometric pattern with a carrier spatial frequency, which does not exceed ~40 mm^{-1} (the resolving power of the sensors).

As an example, let us perform the engineering calculations for a wind tunnel having apertures of Schlieren mirrors ≈200 mm in diameter with focal lengths of ~1,800 mm. If the grating (12.5 mm^{-1}) is located 800 mm from the diaphragm (13), then the size of a light spot on the grating is equal to ~89 mm.

The convergent beam reflected from the flat mirror (11) reads $89 \times 5 = 445$ vertical fringes from the grating (12). A doubled quantity is imaged on a sensor, due to interference of first orders: $445 \times 2 = 890$ fringes. If the spot of light (532 nm, the second harmonics of a Nd:YAG laser) on the grating is imaged on the CMOS sensor (Alpha-900) in a circle 24 mm in diameter, the frequency of the carrier fringe is ~37 mm^{-1}. As to the gratings with a low spatial frequency, the author recommends amplitude ones (Ronchi Ruling Chrome on Glass) manufactured by Applied Image Inc., Rochester, NY, USA. Recall that the spatial resolution of the sensor is ~42 mm^{-1}. Thus, incorrect location of the grating or its excessive spatial frequency may lead to the carrier spatial frequencies of holograms imaged on a CCD/CMOS sensor exceeding its resolving power. The calculations show that the distance between the two circles of confusion on the diaphragm (13) is equal to ~4.25 mm.

Spacing between the two sheared beams, which expose the sensor (15), can be easily changed by a translation moving the grating (12). The coefficient of magnification of an aerodynamic phase object on the sensor is $M = 24/200 = 0.12$. The maximal theoretical spatial resolution in the field of visualization is ~1 mm.

7.4 Applications of Mach–Zehnder Interferometry Using CCD/CMOS Sensors

7.4.1 Vaporization of Volatile Fuel Droplets in a Turbulent Chamber

The homogenous formation of reactive gaseous mixtures in liquid fuel combustion chambers depends largely on the fuel droplet vaporization process.[111,112] The study addresses the particular problem of turbulent effects of droplet vaporization. A turbulence chamber producing zero-mean velocity, isotropic, and homogeneous turbulence is used to investigate the proper influence of turbulence on the vaporization processes of several types of liquid volatile droplets under normal temperature and pressure conditions.

The experimental setup is shown schematically in Figure 7.3. A turbulent chamber (6) in the shape of a cube with external dimensions of $400 \times 400 \times 400$ mm^3, was equipped with three windows to allow access to the chamber both materially and optically. Eight electric fans are fixed at the corners.

The rotation speed of the fans can be regulated from 750 rpm to 1,800 rpm. A quartz fiber is introduced into the chamber from the top at the center of the chamber. The fiber is 0.2 mm in diameter and has an enlarged extremity to ease droplet suspension. A single droplet (7) is suspended on the fiber extremity with a retractable syringe.

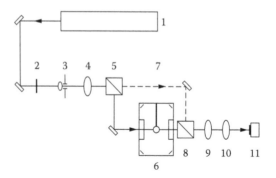

FIGURE 7.3
Focused image Mach–Zehnder interference scheme using a CCD sensor. (1) Helium–neon laser; (2) neutral density filter; (3) spatial filter; (4, 9, 10) objective lenses; (5, 8) beam splitting cubes; (6) turbulent chamber; (7) suspended droplet; (11) CCD-sensor. The dashed line shows the reference beam. 100% reflection aluminum-coated mirrors are not numbered.

The Mach–Zehnder interference scheme was applied to visualize and study the process of droplet vaporization in a turbulent atmosphere of air. A 10-mW helium–neon laser (1) with a wavelength of $\lambda_p = 632.8$ nm was used as the coherent radiation source. The probing light beam was attenuated by a neutral density filter (2), spatially filtered by a spatial filter (3), and expanded by a collimator consisting of a micro-objective of the spatial filter (3) and an objective lens (4). The first beam-splitting cube (5) divides the collimated diagnostic beam on the reflected object and reference beams.

The object beam of the interferometer probes a volatile droplet (7) inside the turbulent chamber through the two viewing windows of high optical quality. A telescope (9, 10) consisting of two objective lenses sharply images the droplet on the sensitive area of a camera-lens-free CCD-sensor (11), which has the resolution of 512(H) × 512(V) pixels.

The evaporating droplet should be imaged with a maximum possible coefficient of magnification, and consequently, spatial resolution. Being attached, the CCD-lens, which is intended for focused imaging of objects located far away from the pupil plane (a long working distance), makes it impossible to achieve a maximal spatial resolution. From this point of view, it is more preferable to use the telescope (9, 10). However, in the telescope the second objective lens used can be the original CCD-lens from the CCD-camera.

A typical microinterferogram with a relatively high-spatial frequency of $f \approx 27$ lines × mm^{-1} (~0.145 lines × pixel^{-1}) for the background fringe pattern is presented in Figure 7.4. The CCD sensor catches ~74 interference fringes. The droplet is imaged at a magnification of 8.7×. The optical scheme provides the theoretical spatial resolution of the order of ~1.25 µm. Due to uncompensated aberrations, the real spatial resolution is 3 to 4 µm.

The resulting fringe shifts are too small to be found by visual standard techniques.[8] It is obvious that the vapor–air mixture over the droplet is the

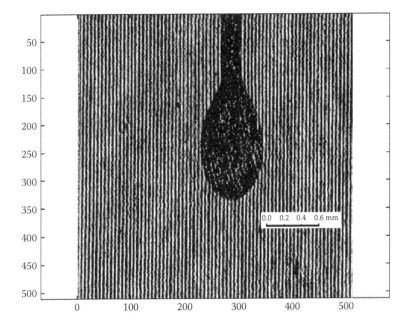

FIGURE 7.4
CCD-interferogram of a vaporizing chloroform droplet. The rotation speed of fans is 1,800 rpm.

thin phase object, and due to the uncompensated interference aberrations, the useful phase shifts probably cannot be measured. Detailed analysis of the situation is presented in Section 3.2. On the other hand, numerical data characterizing the turbulence of the mixture at short distances from the droplet could be found from the contrast measurements of interference fringes.

7.4.2 Laser Breakdown of Tap and Pure Water

As was demonstrated in Section 1.3, the optical breakdown in tap water is realized on inclusion particles. Accurate analysis shows that a laser spark column in reality consists of a train of the phase objects accompanying the solitary thermo-explosions of an individual inclusion subjected to optical "micro-breakdown." At the shortest time delays (a few ns), the laser spark column in tap water is filled with a train of small luminous plasma micro-balls about a hundred microns in diameter, as can be seen in Figure 7.5(a,b). A dozen nanoseconds later, microbubbles appear as the result of an expansion of water vapors surrounding the inclusion particle. Walls of the expanding bubble generate the third objects in a laser spark column: powerful microspherical shock waves over the exploding inclusion. For longer time delays, spherical shock waves over solitary inclusion particles interfere in a

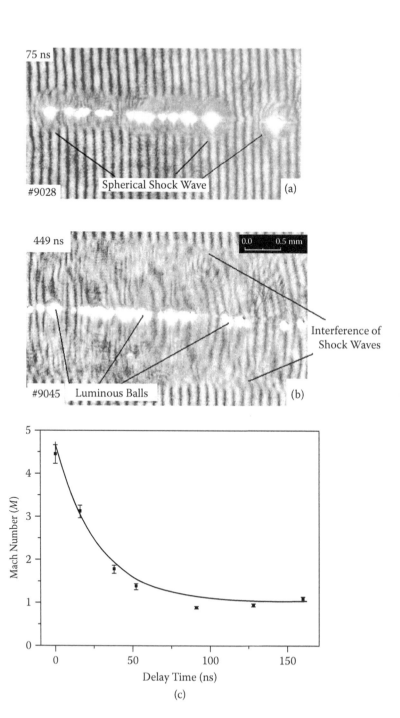

FIGURE 7.5
CMOS interferograms (a, b) and dynamics (c) of the 1,064 nm breakdown in tap water.

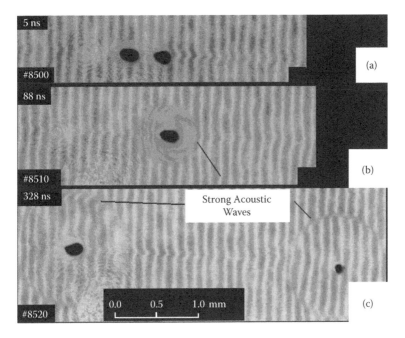

FIGURE 7.6
CMOS interferograms of 1,064 nm breakdown in pure water.

"cylindrical" strong acoustic wave. The average dynamics of the shock waves at time delays shorter than 150 ns is presented in Figure 7.5(c), where the Mach number of the shock waves is shown as a function of the time delay.

The structure of a laser spark column in tap water differs from pure water only numerically; the quantity of luminous microballs in a train of tap water is about one dozen to two dozen in the focal volume of the focusing singlet. In pure water, a quantity of inclusion particles is strongly restricted in comparison with the case of tap water. This speculation is confirmed by the series of interferograms imaged in Figure 7.6, where the dynamics of the region of breakdown is presented. It is seen that the laser spark columns may consist of two or even of one micro-breakdown. The structure of a laser spark column in tap water and ethanol was investigated in Reference [7] by Mach–Zehnder interferometry with imaging of the plasma objects on the CMOS sensor of the Alpha-900 CCD camera.

8

Related Diagnostic Techniques

8.1 White Light Source and Laser Schlieren Diagnostic Techniques for Wind Tunnel Testing by Using CCD/CMOS Imaging

In Section I, problems related to the study of phase objects by applying Mach–Zehnder optical classic and holographic interferometry were discussed. In regard to the holographic approach, the reconstruction schemes of studying a signal wave by using holographic Schlieren and shadow techniques were analyzed. In this section the standard Schlieren techniques will be presented by applying a white light and laser sources.

8.1.1 White Light Source Schlieren Technique

Up to the present time, the classic white light source Schlieren technique remains one of the most simple and popular testing methods of compressible flows in aerodynamics laboratories. The basic concept of the Schlieren technique is to record the deflections of diagnostic beams caused by the compressible flow under test. Schlieren photography is the most frequently applied optical visualization technique because it combines a relatively simple optical arrangement with a high degree of spatial resolution and mechanical stability. The Schlieren approach is used mainly to obtain qualitative information. It is especially applicable for the first steps of visualization and studying the phase objects in powerful aerodynamics, plasma physics, and other large-scale facilities. The base principles and methods of Schlieren photography are illustrated in References [113–116].

Qualitative Schlieren photographs and CCD-focused images can be recorded by using an original white light source Schlieren optical scheme[117–118] presented in Figure 8.1. In the presented setup, a high-pressure mercury lamp with characteristic size of the plasma arc of 0.2×0.2 mm^2 is used as the white light source (1).

This is an example of the simplest and most reliable scheme, which is called the single-pass Z-configuration with the two spherical Schlieren mirrors (3, 7). The Z-configuration Schlieren system is not free of optical

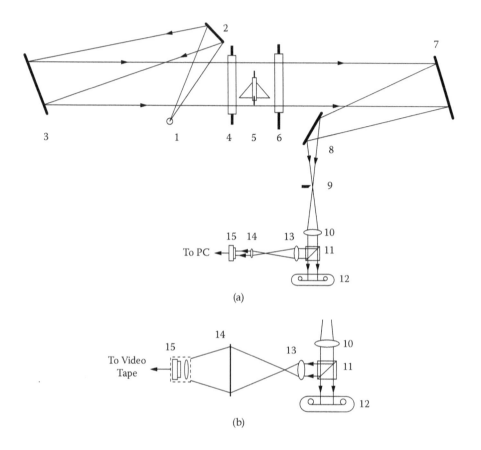

FIGURE 8.1

(a) Two-receiver white light source Schlieren scheme. (1) White light source; (2) flat mirror; (3) spherical Schlieren mirrors (long focal length); (4) window; (5) model; (6) window; (7) spherical Schlieren mirror; (8) flat mirror; (9) optical knife; (10) objective lens; (11) beam-splitter; (12) photographic film; (13, 14) collimator; (15) CCD sensor. (b) Optical channel of the imaging of a flow field on a screen (14) and recording by the CCD camera (15). Optical elements of the second option's scheme: (11) beam-splitting cube; (13) objective lens; (14) screen; (15) CCD camera.

aberrations, due to a large aperture (spherical aberrations) and a relatively large angle between the divergent (converging) beams and a large-aperture collimated diagnostic beam that penetrates the flow field over the model (5) under study. On the other hand, the aberrations of a laser bean enlarge the effective diameter of the circle of confusion in the back focal plane of the Schlieren mirror (7). A larger circle of confusion allows an easier adjustment of the optical knife (9) in the vertical direction, responsible for the quality of a Schlieren picture in the image plans: on the photographic film (12) and the charge coupled device (CCD) sensor (15). Contrary to the solitary receiver module in conventional Schlieren systems, the discussed optical scheme has two receiver modules, which were designed in the Supersonic Wind Tunnel Lab at the Technion. This was achieved due to the additional optical element, the beam-splitting cube (11), which makes it possible to simultaneously record shock flows under test on both photo film (12) and a CCD/CMOS sensor (15). The proposed two-receiver Schlieren system has some advantages over conventional Schlieren devices. It allows optical information to be gathered simultaneously in the two channels. In the first channel, the model (5) is imaged by the cofocal optical system consisting of the Schlieren mirror (7) and the collimating and imaging objective lens (10), which provides sharp focusing of the flow field over the aerodynamic model (5) on a 35-mm photographic film (12). The second channel consists of the beam-splitting cube (11) and the collimator (13, 14), which images the flow field on the sensitive area of a CCD sensor (15). The sensor is a part of a television-type CCD camera without its original lens, included in the collimator, and is used to transfer the image to the hard disk of a PC (see Figure 8.1[a]). Thus Schlieren pictures can be recorded in the regime of movie-Schlierengraphy. In the second option, as is shown in Figure 8.1(b), the lens (14) could be replaced by a large-aperture white screen, and the phase object can be imaged on the screen by the lens (13); after that the image is captured by the CCD camera (15) with the TV-CCD lens on it. The Schlieren images could be transferred in the regime of real time to a magnetic video tape recorder.

A series of white light source Schlieren photographs are shown in Figure 8.2. The model under investigation was a sharp 10-deg 2D wedge (200-mm long and 300-mm wide) with a rectangular cavity (25-mm long). The model includes a narrow, 1-mm-wide slot on the front wall of the cavity in the lower left corner. Through this slot, it was possible to inject a flat jet of air with pressure levels of 0 to 12 atm. The Mach number of the undisturbed flow was $M = 3.4$. Figure 8.2(a–c) demonstrates the effect of injection at the pressure levels of 0, 2, and 10 atm, respectively.

The two-dimensional nature of the wedge makes the spatial resolution of the field of view rather low. Due to a lower air density level in the cavity, it is difficult to get information on the structure of the flow in this area. Nevertheless, it is possible to see some details in the region of reattachment, near the upper right corner of the cavity. In the case of a high jet pressure (Figure 8.2[c]), the shock wave that is generated in the supersonic external flow is clearly visible.

8.1.2 Laser Schlieren Technique

The Schlieren system discussed above may be used with a continuous wave (CW) or pulse laser instead of the white light source. The two-receiver channel laser Schlieren optical scheme[118] is presented in Figure 8.3, where a continuous wave argon-ion laser (1) replaces the white

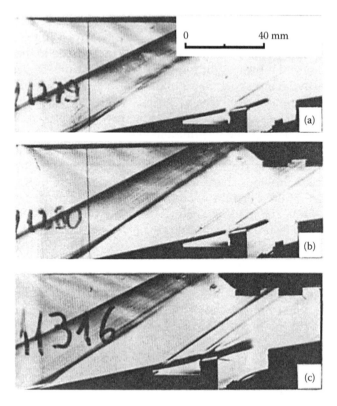

FIGURE 8.2
Schlieren photographs of a shock wave flow over 2D wedge model using a white light source. Jet pressures: (a) 0, (b) 2, and (c) 10 atm. The time-exposure is ~2 ms.

light source. The argon-ion laser can effectively operate at two separate powerful lines with wavelengths of 488.0 and 514.5 nm, land on output power of the order of ~60 mW.

The spherical wave that overfills an aperture of the first spherical Schlieren mirror (6) is generated by the spatial filter (4), which contains a pinhole placed in the focal plane of the spherical mirror. The Schlieren knife (12) is used in the same way as it was used in the white light Schlieren system, i.e., as a spatial diaphragm. The laser Schlieren technique is applicable only because of a remarkable value of spherical, astigmatic, and comatic aberrations. The aberrations increase the size of the circle of confusion in the focal plane of the Schlieren mirror (10) that allows the optical knife to be easily adjusted and quality Schlieren patterns to be obtained in image planes of the collimating objective lenses (13) and (17).

The flow field over the wedge model was recorded by using the two-receiver laser Schlieren scheme and the holographic shearing interferometry scheme (Figure 4.1) for comparison. The laser Schlieren picture (Figure 8.4[a])

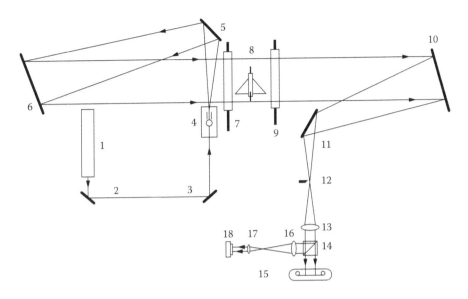

FIGURE 8.3
The two-receiver channel laser Schlieren scheme. (1) Argon-ion laser; (2, 3) flat mirrors; (4) spatial filter; (5) flat mirror; (6) spherical mirror; (7) window; (8) model; (9) window; (10) spherical Schlieren mirror; (11) flat mirror; (12) knife-edge; (13) objective lens; (14) beam-splitting cube; (15) film; (16, 17) collimator; (18) CCD/CMOS sensor.

was obtained when the jet's pressure was equal to zero. Various details of the flow field are seen better than in the case of the white light source Schlieren photographs. For example, the borders of the boundary layers are more defined by the thicker layers. For comparison, Figure 8.4(b,c) illustrates the two reconstructed shearing interferograms of the same phase object. One may see that the information on the structure of the flow field, which is given by the interferograms, is very poor. Most areas of the flow are strongly turbulent, and the interference fringes are not easily seen in these areas. Finally, note that in comparison with the white light source, the laser Schlieren system produces pictures with more contrast and finer structural details for the phase objects with relatively large gradients of the refractive index in the direction perpendicular to the knife's edge. However, for lower pressures—for example, in hypersonic flows with relatively small absolute values of gradients in the flow field—the white light source Schlieren technique may show better results.

In Reference [119] the author has compared the characteristics of a white light source Schlieren with holographic Schlieren techniques. It turned out that diffraction effects are more predominant in the case of the laser illumination. One clear difference can be seen in turbulent regions, where the extremely short-duration laser pulse stops the flow to the extent that fine structures can be seen. After the knife's edge passes the focal point of the Schlieren system, sensitivity decreases, and only a strong phase variation is visible.

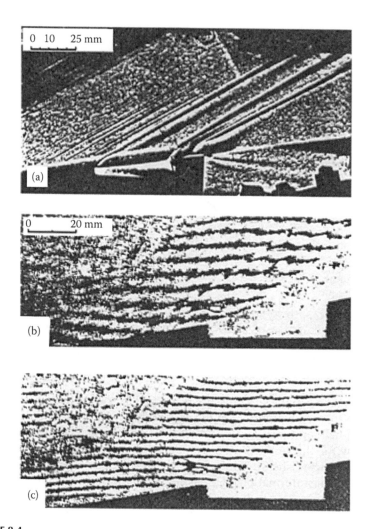

FIGURE 8.4

The shock flow over the 2D wedge model. (a) CW laser Schlieren photograph; (b, c) reconstructed shearing interferograms. The time-exposure is ~2 ms. Schlierengram and interferograms were recorded on a photographic film. Interferograms have two different types of background fringes.

A comparison of laser-Schlieren and hologram-Schlieren images of a bow shock at low pressures was performed in Reference [120]. The authors discovered that the contrast of the holographic Schlieren image is not as high as the laser Schlieren image. The directly recorded Schlieren images are of much higher applicability to understanding complex flow phenomena than holographic reconstructed Schlieren pictures.

It should be taken into consideration that optical information recorded on a flow hologram is more important, and as a full-scale study product than information obtained by a white light source or laser Schlieren techniques. This is connected to the diagnostic opportunities which are offered by the recording of optical information on the hologram. Having only one signal hologram of the phase object under study, one may reconstruct a series of Schlieren images with different positions of an optical knife (in the scheme of reconstruction by a CW laser). Moreover, the optical knife can be replaced with different spatial diaphragms, and finally, the flow field can be visualized and measured by all the applicable techniques described in Section I.

It seems that flow holograms could be reconstructed by light sources with the restrained length of coherence: by an LED source, for example, that has a relatively narrow bandwidth on the order of 10 nm, or by a narrow-bandwidth white light source. A mercury arc lamp, which was used as the light source in Reference [121], illuminates an object by using an interference filter, which isolates the mercury green light at 546.1 nm. The problem is that a laser light (for example, the reconstructing light of a helium–neon laser) has much better coherent characteristics compared with a color LED light or isolated line of a white light source. The coherent light may result in stray interference and speckle structures in the image plane, deteriorating the quality and the spatial resolution of an image.

Finally, a few words in conclusion to be said about the three most applicable light sources for Schlieren applications: an LED, a laser, and a white light source. They may differ not only in spectral (working optical range) characteristics, but also in temporal characteristics. One of the principal shortcomings of such sources is a long time exposure during recording the flow under investigation. A long time exposure may average the quick-changing flow fields. Short pulses of white light and LED sources, and especially of pulsed laser systems, may "freeze" gas dynamic characteristics of the flow field, such as density and velocity, facilitating the study of transient phenomena.

8.1.3 Laser-Induced Spark Schlieren Imaging

The transmission of light beams through compressible flow fields in a wind tunnel with index-of-refraction variations recently has been extensively studied by different Schlieren visualization techniques.[122] Most of the published experimental works used a diagnostic beam transmitted through the flow field in the direction perpendicular to the flow axis. Measuring the flow field in the direction of the stream is a very complicated problem, because it is impossible to insert a light source into the facility without disturbing the flow.

The Schlieren system in which a white light source is inserted into the flow upstream of the model without disturbing its parameters was discussed in Reference [123]. The experimental concept of such a system is shown in Figure 8.5(a). The suggested pulsed white light source is a laser spark in air

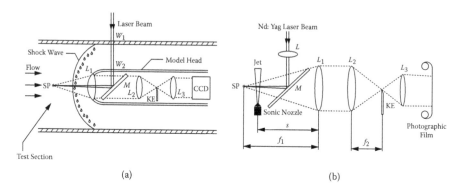

FIGURE 8.5
(a) Proposed and (b) experimental "laser spark" Schlieren systems.

initiated by a Q-switched Nd:YAG laser with parameters of $\lambda = 1.064$ μm, $E = 300$ mJ, and $\tau_{\frac{1}{2}} = 10$ ns.

Lens L_1 focuses the laser light and at the same time serves as a collimating lens for the laser spark as a pulsed white light source. A telescope (L_2, L_3) images the shock flow on the sensitive area of a CCD camera. A knife-edge is located in the back focal plane of the "Schlieren" lens L_2.

A simplified experimental system is presented in Figure 8.5(b). It is seen that in the present system the phase object is placed in the diverging beam at distance s from L_1, rather than in the parallel beam region as in conventional Schlieren systems.

For the paraxial approximation, the displacement of the spark image δ, because of the angular deflection of the optical ray caused by the phase object, is given by:

$$\delta = (F_1 - s) \times (F_2/F_1) \tag{8.1}$$

This equation reveals that the sensitivity of the system increases when the distance s is shortened and the ratio F_2/F_1 is increased.

A typical "laser spark" Schlieren photograph of a free jet (the jet is described in detail in Reference [3]) is shown in Figure 8.6(a). For comparison, a Schlieren photograph without the jet is demonstrated in Figure 8.6(b). The system is of a high spatial resolution, because the turbulent density structures of the order of 0.1 mm are clearly resolved. Schlieren images of the jet taken with consequent laser pulses were highly reproducible.

Note that a short-duration white light source, ~0.2 μs, of the Schlieren system allows photographing instantaneous density fields. Moreover, a laser spark in air initiated by a Q-switched laser pulse could be an effective very short (≤1 microsecond) white light source for conventional Schlieren devices; therefore it may be located in the required area of a facility.

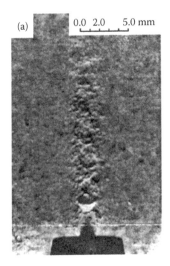

FIGURE 8.6
"Laser spark" Schlieren pictures of supersonic micro jet. (a) Schlieren picture of a free micro jet; (b) Schlieren picture without the object. The initial size of the jet is 1.3 mm.

8.2 Application of Moiré Deflectometry Technique for Gas Dynamic Research

One of the first approaches regarding successful application of Moiré deflectometry to mapping an axisymmetric shock flow in a wind tunnel has been suggested in Reference [124]. The principal optical elements of any deflectometry scheme are two identical Ronchi (amplitude) gratings having a pitch p, which are set a distance Δz apart and oriented at angles $+\theta/2$ and $-\theta/2$, respectively, relative to the y-axis. When a collimated beam passes through the gratings, a Moiré pattern is produced in the imagining plane optically conjugated to the second grating. The undisturbed pattern on the second grating consists of straight fringes parallel to the x-axis separated by a distance $p^* = p/[2 \sin (\theta/2)]$. Indeed, if the gratings were rotated at the angle $\pm\theta/2$, then the resulting vector of spatial frequencies of the background Moiré pattern is $|f_{res}| = 2 \times |f| \times \sin(\theta/2)$, where $|f| = 1/p$ and $|f_{res}| = 1/p^*$. Thus rotating the amplitude gratings could change the spacing of the background Moiré pattern. If $\theta \approx 0$, then the width of a Moiré fringe exceeds the size of the field of view, and the background Moiré pattern is called an infinite-fringe Moiré deflectogram. In the first approximation, the lateral shift s between the zero- and first-order beams just behind the first grating is determined as $s \approx \Delta z \times \lambda/p$, where λ is the wavelength of a diagnostic laser and under the condition that $\lambda/p \ll 1$.

When the signal collimated beam propagates through a flow field, it is disturbed due to density gradients and the Moiré pattern is deformed. The

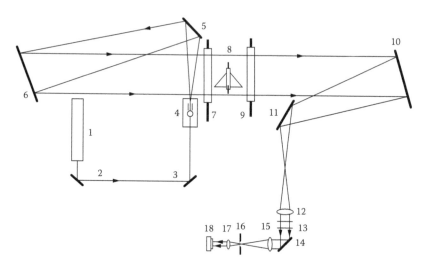

FIGURE 8.7

Moiré deflectometry scheme with optical filtering of the resulting Moiré picture. (1) Argon-ion laser; (2, 3) flat mirrors; (4) spatial filter; (5) flat mirror; (6) spherical mirror; (7) window; (8) model; (9) window; (10) spherical mirror; (11) flat mirror; (12) objective lens (13) amplitude gratings; (14) flat mirror; (15, 17) collimator; (16) stop diaphragm; (18) CCD/CMOS sensor or photographic film.

calculations of the beam deflection angles using controllable fringe shifts on a Moiré interferogram make it possible to determine the refraction index gradients of the perpendicular to Moiré fringes. Details of the Moiré deflectometry technique are discussed in Reference [125].

A novel optical Moiré deflectometry scheme is illustrated in Figure 8.7. The focused image scheme could be successfully applied to compressible wind tunnel testing and may consist of the same optical elements from which the white light and laser Schlieren schemes discussed in the previous paragraph were designed. The principle difference is the design of the receiver modules.

The optical knife is removed, and the collimating objective (12) provides the focusing of a phase object in the plane of the second amplitude grating (13). The collimator (15, 17) provides optical conjugation of the second amplitude grating on the sensitive area of the CCD camera (18) (the CCD lens is detached from the camera). In Figure 8.7 the second receiver channel with a photographic film is not shown.

It is worth remarking on some positive characteristics of the proposed scheme in comparison with conventional setups. First, conventional Moiré deflectometry schemes, suggested for aerodynamic flow field and other investigations in References [126–129], operate with collimating or spherical beams, where the second Ronchi grating is not focused in the image plane. The sharp imaging of a phase object onto photographic film or on a CCD/

CMOS sensor greatly enhances the spatial resolution of the image and allows more accurate retrieval of Moiré deflectometry data. Second, the use of the telescope system (15, 17) makes it possible to realize the optical filtering in the back focal plane of the lens (15) by means of a stop diaphragm (16). Only one order (usually the brightest) is selected and focused in the imaging plane; all other orders are blocked out. In the results, due to the procedure of optical filtering, the experimenter may achieve greater spatial resolution of the imaged object and improve the contrast of the Moiré pattern. Without the filtering, the resolution and contrast are reduced because of the effect of multibeam interference of overlapped diffraction orders behind the second grating. Note Reference [129], where the method of holographic reconstruction of Moiré deflectograms is suggested for the first time (see Section I of this book).

Typical deflectograms of a turbulent subsonic heated jet are shown in Figure 8.8 (compare with the reconstructed shearing interferograms

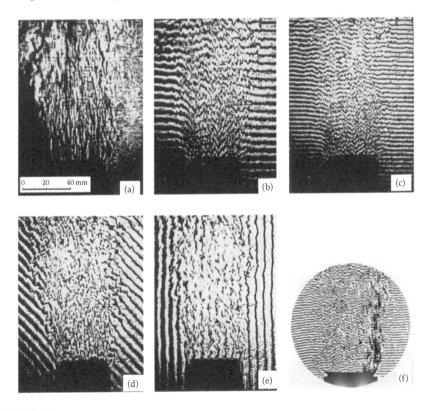

FIGURE 8.8
Moiré deflectograms of subsonic turbulent heated air jet and a reconstructed shearing interferogram. (a) Infinite- and (b–e) finite-width-fringe Moiré deflectograms; (f) reconstructed finite-width-fringe shearing interferogram. The fields of view: (a)–(e) 100 × 136 mm^2; and (f) diameter 100 mm.

presented in Figure 5.1). It is seen that the shearing interferograms, including the one shown in Figure 8.8(f), demonstrate higher spatial resolution and better quality of background fringes. Because the same optical elements were used in both optical schemes, the irremovable effect of the presence of higher diffraction orders in the selected orders is responsible for deteriorating the spatial resolution and quality of the Moiré pictures.

9

Devices and Equipment for Routine Mach–Zehnder Holographic Interference Experiments

9.1 Methods of Spatial Filtering Applicable to Mach–Zehnder Holographic Interference Experiments

9.1.1 Spatial Filtering for Schemes of Recording and Reconstruction

Spatial filtering methods are strongly connected to characteristics of holographic experimental schemes designed for recording and reconstructing Mach–Zehnder holograms. They accompany practically all stages of applications of the holographic experiment, mainly because of low spatial carrier frequencies of the holograms. The optical filtering is increasingly effective due to the use of one- or two-hole diaphragms, which select one or two required reconstructed waves; all others are blocked out. It should be noted that proper selection of a coherent laser source, a focusing objective lens, and a filtering diaphragm with the required diameters of the holes and the distance between them are the only guarantee of quality optical filtering. A coherent continuous wave (CW) laser source is necessary for obtaining a minimal diameter of the circle of confusion in the back focal plane of the focusing objective lens. The CW laser is connected with a little distance between correspondent diffraction orders in the back focal plane of the objective lens. The same circumstance is responsible for a long enough focal length of the focusing objective lens. A normal inciting collimated beam behind a Mach–Zehnder hologram with the spatial frequency of carrier fringes f that is focused by a collimating objective lens generates diffraction orders with the angular distance between zero and first order: $\sin(\alpha) = f \times \lambda$, where λ is the wavelength of a diagnostic beam. The linear distance, s, between any two nearest circles of confusion in the focal plane is expressed [$\tan(\alpha) \approx \sin(\alpha) \approx \alpha$], in accordance with Equation 2.1. For example, for $f = 25$ mm^{-1}, the spacing between zero and first order will be $s \sim 2.57$ mm, if $F = 200$ mm and $\lambda = 0.5415 \times 10^{-3}$ mm.

The diameter of the hole of the diaphragm should be a little larger than an effective size of the circle of confusion; otherwise the light from the neighboring diffraction orders could distort the selected beam.

A one-hole stop diaphragm (or an optical adjustable slit) is used in the following cases:

1. Selection of a signal wave recorded on the signal hologram during the process of reconstruction (see Figure 2.2 and Figure 2.3[b]) for the purposes of testing it by holographic Schlieren, shadow, Moiré deflectometry, and dual reference hologram shearing interferometry techniques

2. Selection of signal and comparison waves simultaneously, being recorded on a double exposure hologram during the process of reconstruction (see Figure 2.7) or during the process of reconstruction of the same waves recorded on separate films (dual hologram technique, Figure 2.9)

3. Selection of existing real-time signal and comparison wave, being recorded on a comparison hologram, during the real-time interference experiment (see Figure 2.10)

4. Selection of a signal wave reconstructed from a double exposure hologram for the purpose of testing it by the shadow technique and possibly by the Schlieren technique

5. Selection of only one reconstructed order in the scheme of a real-time holographic shearing interference technique (see Figure 6.1[b])

6. Selection of only one diffracted order in schemes of Moiré deflectometry (see Figure 8.7).

A two-hole stop diaphragm is used in the following cases:

1. Selection of $\pm m$th diffraction orders reconstructed from a master hologram, for the purpose of recording the secondary signal and comparison holograms (see Figure 3.8), in frames of reference beam and shearing holographic interferometry with the enhanced sensitivity approach

2, Selection of ±1st diffraction orders of signal/comparison waves (double exposure, dual hologram, and real-time holographic shearing interferometry) to record the signal/comparison shearing hologram (see Figure 4.1 and Figure 6.1[a])

9.1.2   Cleaning the Diagnostic Beam of a CW Laser

The concept of cleaning a diagnostic laser beam is applicable to a narrow pencil of a continuous wave laser. CW argon-ion lasers (in this case, the spatial cleaning must be performed more carefully, because the pinhole could be damaged by the focused beam ejected from the argon-ion laser at elevated powers >100 mW) and helium–neon and other types of CW lasers working in the visible range have the characteristic sizes of beams of the order of 0.8 to 1.5 mm with divergences of the order of ~1 × 10^{-3} to 1.5 × 10^{-3} radians. For small diameters of the beams, the light scattering from small dust particles and scratches on the surfaces of optical elements is the critical issue. As a result of these effects, additional interference noise and scattering appear, and the optical quality of the pencil beam is deteriorated. In order to avoid these effects, the technique of spatial cleaning is applicable in the middle of holographic experiments by means of a so-called "spatial filter." For this purpose, the narrow pencil laser beam is focused by a high-quality microscopic singlet in the plane of a pinhole diaphragm with diameters of the order of ~10 to 20 μm, which provides the required optical filtering.

It is well known[130] that Fraunhofer diffraction of the plane wave on a circular aperture of a high-quality microscopic lens consists of a bright central spot with the following diameter:

$$d_A = 2.44 \times \lambda \times F_\# \tag{9.1}$$

and numerous surrounding rings, with 83.8% of the laser energy in the central spot known as the Airy disk, and $F_\#$ *(F-number)* defined as the ratio of the focal distance to the aperture of the lens.

Let a narrow pencil of a He–Ne laser be focused by the micro-objective with $F_\# = 10$. Then the diameter of the central spot is equal to $d_A \approx 2.44 \times 10 \times 0.6328$ μm, ~15 μm. The latter value means that the diameter of the proper pinhole diaphragm should be chosen to be approximately ~15 μm. The real diameter of the correct pinhole diaphragm is selected experimentally. Practically, it should be of the order of the diameter of the Airy disk. The Airy disk is formed with the lowest spatial frequencies of the collimated beam.[130] The micro-objective lens performs the Fourier transform of the collimated diagnostic beam in the focal plane.[9,130] The unwanted light, owing to the interference noise and scattering, has higher spatial frequencies and is focused in the back focal plane of the microlens, far from the Airy disk. Thus the unwanted light is cut by the pinhole. The light selected by the pinhole has a very smooth irradiance distribution due to the lowest spatial frequencies, and will propagate in the form of an almost ideal spherical wave, which may be collimated by a high-quality objective lens for further use in an interference optical scheme. Note that the pinhole diaphragm should be carefully adjusted in the (x, y) plane and in the z direction by means of microscopic screws during the procedure of filtering.

9.2 Devices and Equipment Required for Performing Routine Mach–Zehnder Holographic Interference Experiments

In this section, basic devices, equipment, and optical and mechanical elements necessary for performing Mach–Zehnder holographic interference experiments are briefly discussed. Coherent laser sources and optical elements necessary for designing a holographic interferometer's optical schemes are also analyzed.

Lasers for holographic interferometry are divided into two main groups: *continuous wave* (CW) and *pulsed laser* systems. The mode structure of the chosen laser system should be as good as possible. The temporal coherence of the laser light, which is responsible for the length of coherence l_c, can be evaluated from the following simple expression: $l_c \approx \lambda^2/(\Delta\lambda) = c/\Delta v$, where λ is the wavelength, $\Delta\lambda$ is the width of the laser line, c is the velocity of light, and Δv is the bandwidth of the laser line: $\Delta v = c\Delta\lambda/\lambda^2$. The length of coherence determines the maximum possible optical path difference between the object and reference arms of the designed interferometer. If this value increases, the contrast of a recorded hologram and its diffraction efficiency deteriorate, finally leading to a low intensity of the reconstructed waves.

Most popular CW coherent sources are argon-ion and helium–neon laser systems. The coherence length for an ordinary He–Ne laser, having the line width of the order of ~150 MHz and $\lambda = 632.8$ nm, is equal to $l_c \approx 20$ cm. Standard He–Ne lasers generate about 2 to 10 mW of output light power and are well suited to Mach–Zehnder holographic interference experiments. The wavelength $\lambda = 632.8$ nm requires panchromatic photo materials. A CW krypton-ion laser working at the wavelength of 647 nm, in the red region of visible light, could also be useful for interference experiments.

Argon-ion gas laser systems are characterized with relatively broad laser lines of about 3,500 MHz and a coherence length of ~8 cm. This is shorter than the coherence length of a He–Ne laser and requires more careful handling and adjustment of the interferometer arms through the use of an argon laser. On the other hand, the argon-ion laser generates two more powerful lines in the blue-green region of the visible range (i.e., 514.5 and 488.0 nm) required for recording on orthochromatic photo materials. The second popular laser in the blue range is a helium–cadmium laser with the working wavelength of 442 nm. Its violet line also requires orthochromatic photo materials.

The two types of pulsed lasers that are most popular in holographic interference experiments are ruby ($\lambda = 694.3$ nm) and frequency-doubled Nd:YAG laser ($\lambda = 532$ nm) systems. These lasers are described in detail in Reference [131], which is dedicated to the questions of using pulsed laser systems for holographic interferometry. The lasers operate in a Q-switched regime that is characterized by a short 5- to 30-ns time duration and powerful pulses (dozens of megawatts) of generation. The Q-switched regime operates in passive or active modes. The active mode allows a laser pulse to be synchronized with the arising phase object or with some time phase of its evolution. Here, the accuracy of the synchronization could be very high and compared with the laser pulse length. In the situation of passive Q-switching, the degree of synchronization is worse by a few microseconds.

Conventional ruby lasers have coherence lengths on the order of ~2 cm. For the purpose of enhancing the length of coherence, a special optical element—the etalon—is placed in the resonator. An etalon is manufactured from high-quality, parallel fused silica, quartz, or sapphire plates. It effectively selects longitudinal modes, making the line of generation much narrower, and thus enlarging the coherence length up to ~100 cm. The best systems generate on the order of ~0.1 to 1.0 J of output light energy, which is enough for recording phase holograms on a large-aperture holographic photo plate. Unfortunately, most of the existing photo materials, including holographic photo materials, are orthochromatic, which are not sensitive to the radiation of a ruby laser line with λ = 694.3 nm, or have a very low sensitivity in the red region. In this case, panchromatic photo materials are sensitized for the purpose of extending their sensitivity in the far red region.

Nd:YAG laser systems generate a working laser line of 1.064 μm (i.e., in the near IR region); however, there is a very restricted assortment of photo materials that are sensitive in near IR region. To obtain the output laser light in the visible region, the frequency of the laser line must be doubled. The frequency-doubled Nd:YAG laser systems provide a green line λ = 1,064/2 = 532 nm. Photo industry manufactures produce a wide range of orthochromatic photo materials, which can be used in Mach–Zehnder experiments, including holographic photo films and plates, and which are applicable for the standard holographic approach.

Additional useful information on physical characteristics, principles of generation, and parameters of the laser systems, part of which is applicable for holographic interference experiments, can be found in Reference [132].

As already reported in Section I, for Mach–Zehnder interferometric and holographic approaches, experimenters may use a huge variety of standard photographic 35-mm films, including panchromatic as well as orthochromatic types.

The type of *mirrors* used depends on their place in the interferometer's scheme, and whether the scheme employs a narrow pencil beam or a wide aperture collimated beam. For a powerful narrow pulsed laser beam, mirrors should resist a high level of damaging laser intensities. In this case, mirrors with an interference dielectric coating must be used (the so-called laser mirrors). Sometimes it is convenient to use the laser mirrors with the highest coefficient of reflection at 45-deg angles. The mirrors working in wide collimated beams can be coated with aluminum, though it should be remarked that this type of coating is not durable to high laser intensities.

The best *beam-splitters* are cubes; unfortunately they do not resist high light intensities of pulsed laser beams and should be located only in collimated beams.

Objective lenses are an essential part of the optical scheme, and are used for the purpose of filtering, focusing, collimating, and imaging. As a rule objective lenses are resistant to collimated beams of pulsed lasers. Diverging (negative) lenses are durable and are used for expanding narrow intensive pulsed laser beams. Positive lenses are used only for CW lasers because of the possible effect of a laser spark, which could arise in the focal plane of the lens owing to the powerful pulse of a pulsed laser.

Spatial filters, which are used for clearing narrow laser beams of CW lasers, can rarely be used with pulsed systems, because laser sparks could easily damage pinholes of the filter. For pulsed systems a high-intensity durable diaphragm can be used to clean a diagnostic beam in order to choose the most uniform part of the wave front, which is especially important for multimode-pulsed laser systems.

In Section 1.3 it was shown that the selection of applicable photographic materials depends on two factors: (1) the working wavelength of the laser line and (2) the smallest resolvable detail of the phase object under study. Thus, it follows that some standard *photographic materials* with relatively low resolution power, ~100 mm^{-1}, can be successfully used in Mach–Zehnder holographic interference experiments if the resolution requirements are not too hard (>20 μm).

In any case, the experimenter should choose emulsions of the higher contrast (higher gamma) type and resolving power. Small variations of a laser light during the exposure of a hologram lead to relatively large variations in optical densities. Most standard holographic developers, including popular D-19 and D-76, represent contrast working solutions. They are especially designed for development of negatives having high sharpness of fine details. Frequently the developed negatives have extreme contrast areas overexposed (high optical densities).

It is well known that some development procedures may affect the slope of the characteristic curve[15] and change the contrast between light and dark regions. The author offers one of the smoothly working developers, which may reduce the contrast of overexposed areas. The developer is used as a two-component short-life mixture; hence it should be prepared from the two equal parts immediately before the interference experiment. One part of solution A and one part of solution B are added to eight parts water. Table 9.1 provides the recipe for the developer, which modifies the contrast of negatives, removing the overexposed areas.

TABLE 9.1

Smoothly Working Developer

Solution A	Grams
Sodium sulfite, anhydrous	15
4-Methylaminophenol sulfate (Metol)	15
Sodium sulfite, anhydrous	75
H$_2$O (water)	Up to 1 liter
Solution B	Grams
1-Phenil-3-pyrazolidinone (Pyrozolidone)	1
Hydroquinone	15
Sodium sulfite, anhydrous	75
Sodium carbonate, anhydrous	23
H$_2$O (water)	up to 1 liter

Conclusion to Section III

At the present time, digital CCD/CMOS sensors begin to play a determinative role in the procedure of diagnostics of the phase objects under test in aerodynamic and laser laboratories. The only shortcoming of sensors is their relatively low, limited resolving powers. For example, the full-frame CMOS sensors possess resolving powers that do not exceed ~40 mm^{-1}.

In Section III it was shown that reference beam signal and comparison waves recorded on a CCD/CMOS sensor in the scheme of a Mach–Zehnder interferometer could be used not only for obtaining reference beam reconstructed interferograms, but also for digital generation of shearing reconstructed interferograms for the purpose of studying the field of gradients of the refraction index. Recall that the size of the phase object in this case cannot be larger than the linear size of the beam-splitting cubes. As to the digital reconstructed interference pictures of wide-aperture aerodynamic flow fields, they could be recorded in a series of interferometers described in Section 4.1, including the diffracted holographic shearing interferometer, which was analyzed in Section 4.2.

A separate problem is the digital analysis of recording and computer processing of thin phase objects. First of all, it is obvious that such class of objects should be studied exclusively by holographic techniques, when aberrations of the scheme of recording can be canceled. The problem of digitally enhancing the sensitivity of interference measurements stays open and requires special future analysis.

Separate analysis is required for the problem of digital holographic Schlieren and shadowgraph techniques.

Diagnostic white light Schlieren and laser source Moiré deflectometry techniques described in Section III could be simple and very useful methods for preliminary studies of phase objects, mainly at wide-aperture facilities having a relatively low quality and restricted number of optical elements. The material is supplemented with the analysis of the spatial filtering of a Moiré deflectometry scheme necessary for getting deflectograms of a satisfactory quality.

The space filtration of pencil beams of CW lasers is useful and sometimes necessary for getting quality Mach–Zehnder interferograms and holograms. Correctly applied methods of spatial filtering in schemes of recording interferograms/holograms and reconstruction of holograms may help the researcher in conducting proper holographic experiments.

Conclusion

The methods and special techniques of classic and holographic Mach–Zehnder reference beam and shearing interferometry that were presented in this text may successfully solve diverse physical and engineering problems arising during aerodynamic wind tunnel testing and optical diagnostics of the processes of laser–matter interaction, etc. The digital approach supplements methods of optical holographic interferometry and allows obtaining digital reconstructed interferograms without wet-chemical processing of photo materials.

A Mach–Zehnder approach demonstrates a relatively high spatial resolution ($\approx$10 to 20 µm), depending on spatial and contrast characteristics of photographic films or CCD/CMOS sensors and imaging objective lenses, and a high temporal resolution, which is determined by the nanosecond duration of Q-switched laser pulses. Practically all the phase objects studied by Mach–Zehnder interferometry technique are presented here for the fist time. These interesting phase objects are:

- Microspherical shock waves, strong acoustic waves, and bubbles in liquids from inclusions arising due to the process of optical breakdown
- Heated regions in water and alcohols in the focal volume of a singlet
- Supersonic air micro jets
- Vaporizing volatile liquid droplets in still and turbulent air
- Thermal waves in plastic in the middle of the process of laser drilling
- Strong acoustic waves in ethanol

Mach–Zehnder classic and holographic reference beam interferometry is a versatile and practical method for studying phase objects, which have sizes that are smaller than the apertures of beam-splitting cubes. Both methods allow imaging not only on standard photographic films, but also on the sensitive areas of CCD/CMOS sensors that facilitate advanced digital interference measurment and computer reduction of the interference data.

In the classic technique, a sensor is used for direct recording of interference pictures of the phase objects under study. In the holographic approach, a comparison hologram is recorded together with a signal one, which carries useful phase information on a phase object. The holograms are used for optical or digital subtraction for the purpose of obtaining digitally (optically) reconstructed interferograms, which are free of optical aberrations of the scheme of recording. Later they are processed optically or digitally for the purpose of retrieving the interference data.

Despite the fact that electronic CCD/CMOS sensors and standard photographic films have relatively low spatial resolving powers, they are able to successfully record Mach–Zehnder holographic gratings. Mach–Zehnder signal and comparison holograms allow realization of principle advantages characteristic of the optical holographic technique in comparison with its classic counterpart.

Classic optical schemes, where a phase object is imaged directly on the sensitive area of CCD/CMOS sensors, are pretty simple and applicable for express or preliminary analysis of the main characteristics of phase objects. Holographic methods are more advanced and capable of generating reconstruction interferograms having arbitrary spatial frequency and orientation of background fringes. Due to the canceling optical aberrations in reconstructed interferograms, the optical Mach–Zehnder holographic approach makes it possible to measure thin phase objects by enhancing the sensitivity of interference measurements. It allows obtaining a series of reconstructed interferograms from solitary signal (flow) holograms. Moreover, there are potential opportunities for studying a reconstructed signal wave by applying related diagnostic methods.

The recording of low-spatial-frequency Mach–Zehnder holograms on photographic films may restrict a spatial resolution of the reconstructed interferograms and be an obstacle for measuring phase micro-objects (or fine details of phase objects). At the same time, the Mach–Zehnder approach is pretty flexible from the point of view of enhancing the spatial resolution of fine details of a phase object. Using photographic films with better resolving powers than standard ones allows measuring a phase object with better spatial resolution. The other factor, which may essentially restrict an efficiency of interference measurements of micro-objects with large magnifications, is a relatively large distance between the imaging objective lens and the phase micro-object under study.

The author believes that a Mach–Zehnder holographic reference beam approach, together with applying the technique of rewriting holograms during the procedure of reconstruction, would make it possible to successfully study supersonic (hypersonic) micro jets with diameters smaller than ~0.1 mm and other weak phase gas dynamic micro-objects. One of the challenging gas dynamic objects is a free micro jet expanding in vacuum or a low-pressure gas atmosphere. The second challenging phase object is surface laser plasma expanding in vacuum or a low-pressure gas atmosphere.

Challenging gas dynamic problems also exist for applications of Mach–Zehnder shearing holographic interferometry of hypersonic shock flows at high (>15–20) Mach numbers and reduced gas densities in test sections. The author believes that methods of Mach–Zehnder shearing holographic interferometry with rewriting (rerecording) signal and comparison holograms may successfully solve these problems. The other promising gas dynamic project is applying a Mach–Zehnder shearing approach with reduced sensitivity to studying the structure of strong bow shocks.

The newly developed method of dual hologram Mach–Zehnder shearing interferometry demonstrates its unquestionable advantages for investigation of diverse shock flows in wind and shock tunnels, the running of which is connected with large-scale acoustical disturbances. The application of the acoustically tolerant diffraction shearing interferometer guarantees successful measurements by using CW lasers in the case of facilities responsible for practically irremovable mechanical vibrations. The apertures of the phase objects under investigation could be also practically unlimited. The method is more convenient and flexible than classic shearing interferometry.

The dual hologram shearing (and reference beam) method could be very promising and a powerful tool in applications for turbulent plasmas and nonstationary gas dynamic flows. Measurements of gradients of the refractive index can be supplemented with measurements of the contrast of reconstructed or real-time interference patterns recorded on CCD/CMOS sensors. So far as the contrast is a function of parameters of turbulent flows, the turbulence may be characterized by analyzing contrast of reconstructed and classic digital interferograms.

Besides aerospace objects, numerous phase objects arising due to laser–matter interactions can be studied by applying methods of a Mach–Zehnder holographic approach. That is related to acoustic and shock waves in transparent liquids and solids, where regions behind the head of rarefaction and shock (strong acoustic) waves, as well as electron densities of ablation and laser plasmas in gases and other plasma objects, may be visualized and numerically determined.

Studying very short-time-duration transient phase objects by using the diagnostic visible light of second harmonics of a Nd:YAG picosecond duration pulsed laser system involves very interesting, promising, and at the same time complicated problems of interference. Unfortunately ultra-short pulses of a laser light have a very short length of coherence. For example, a relatively long ~1 picosecond laser pulse has a length of coherence shorter than ~0.3 mm, which seems to make only recording of holographic gratings of the Mach–Zehnder type possible.

There is no doubt that comprehensive and flexible methods of the Mach–Zehnder reference beam and shearing holographic approach will grow in favor with scientists and engineers and will find further application as useful diagnostic tools in future aerospace and laser–matter interaction projects.

References

Section I

1. L. Zehnder, "Ein neuer Interferenzrefraktur," *Z. Instrumentenkd.* 11, 275–285 (1891).
2. L. Mach, "Uber einen Interferenzfractur," *Z. Instrumentenkd.* 12, 89–93 (1892).
3. G. Toker, D. Levin, and J. Stricker, "Dual Hologram Technique with Enhanced Sensitivity for Measurements of Weak Phase Objects," *Experiments in Fluids* 22, 354–357 (1997).
4. J. G. Velásquez-Aguilar, G. Toker, A. Zamudio-Lara, and M. Arias-Estrada, "Visualization of a Supersonic Air Micro Jet by Methods of Dual-Hologram Interferometry," *Experiments in Fluids* 42(6), 863–869 (2007).
5. M. Aleksandrov, G. Kuz'min, A. Prokhorov, and G. Toker, "Experimental Modeling of Thermal Process, Arising on Drilling of Hard Biological Tissues by a Pulsed-Periodic CO_2 Laser," *Laser Physics* 3(4), 918–920 (1993).
6. G. Toker, V. Bulatov, T. Kovalchuk, and I. Schechter, "Micro-Dynamics of Optical Breakdown in Water Induced by Nanosecond Laser Pulses of 1064 nm Wavelength," *Chem. Phys. Letts.* 471, 244 (2009).
7. T. Kovalchuk, G. Toker, V. Bulatov, and I. Schechter, "Laser Breakdown in Alcohols and Water Induced by $\lambda = 1064$ nm Nanosecond Pulses," *Chem. Physics Letters* 500(4–6), 242–250 (2010).
8. C. Vest, *Holographic Interferometry*, J. Wiley & Sons, New York (1979).
9. R. Collier, C. Burckhardt, and L. Lin, *Optical Holography*, Academic Press, New York (1971).
10. R. Brooks, "Low-Angle Holographic Interferometry Using Tri-X Pan Film," *Applied Optics* 6(8), 1418–1419 (1967).
11. G. Toker, D. Levin, and J. Stricker, "Experimental Investigation of Super/ Hypersonic Flow Field Based on a New Approach Using Holographic Optical Techniques," *ICIASF'95*, 10.1–10.7 (1995).
12. O. Bringhdal and A. Lohmann, "Interferograms Are Image Holograms," *J. Opt. Soc. Am.* 58(1), 141–142 (1968).
13. L. Tanner, "Some Applications of Holography in Fluid Mechanics," *J. Scien. Instrum.* 43, 81–83 (1966).
14. A. Kozma, "Analysis of the Film Nonlinearities in Holographic Recording," *Opt. Acta* 15, 527–551 (1968).
15. O. Bringhdal and A. Lomann, "Nonlinear Effects in Holography," *J. Opt. Soc. Am.* 58, 1325–1334 (1968).
16. L. Rosen, "Focused Image Holography with Extended Sources," *Appl. Phys. Letts.* 9(9), 337–339 (1966).

17. G. W. Strocke, "White Light Reconstruction of Holographic Images Using Transmission Holograms Recorded with Conventionally-Focused Images and 'In-Line' Background," *Phys. Letts.* 23(5), 325–327 (1966).
18. P. W. Chan and C. S. Lee, "Holographic Schlieren Investigation of Laser-Induced Plasmas," *Physics Letters A* 62(1), 33–35 (1977).
19. J. E. O'Hare and J. D. Trolinger, "Holographic Color Schlieren," *Applied Optics* 8(10), 2047 (1969).
20. W. Merzkirch, *Flow Visualization*, Academic Press, New York (1974).
21. B. Hannah and W. King, "Extensions of Dual-Plate Holographic Interferometry," *AIAA Journal* 15(5), 725–727 (1977).
22. G. Toker and J. Stricker, "Holographic Study of Suspended Vaporizing Volatile Liquid Droplets in Still Air," *International J. Mass and Heat Transfer* 39(16), 3475–3482 (1996).
23. I. Zeilikovich, A. Lyalikov, and G. Toker, "Visualization of Acoustic Waves in a Dye Solution by Holographic Interferometry," *Sov. Technical Physics Letters* 14(3), 213–214 (1988).
24. I. Zeilikovich, A. Lyalikov, and G. Toker, "Visualization of Acoustic Waves in a Dye Solution by Means of Holographic Interferometry," *Russian Ultrasonic* 18(4), 215–217 (1988).
25. V. Afanaseva, L. Mustafina, and V. Seleznev, "Method of Compensating Aberrations in Holographic Interferometry," *Opt. Spectrosc.* 37(4), 448–449 (1974).
26. A. K. Beketova, A. F. Belozerov, A. N. Berezkin, et al., *Holographic Interferometry of Objects with Phase* (in Russian), Chs. 5, 9, Nauka, Leningrad (1979).
27. J. Trolinger, "Application of Generalized Phase Control During Reconstruction to Flow Visualization Holography," *Applied Optics* 18(6), 766–774 (1979).
28. K. Matsumoto and M. Takashima, "Phase-Difference Amplification of Non-Linear Holography," *J. Opt. Soc. Am.* 60, 30–33 (1970).
29. K. Mustafin, V. Seleznev, and E. Shtyrkov, "Use of the Nonlinear Properties of a Photoemulsion for Enhancing the Sensitivity of Holographic Interferometry," *Opt. Spectrosc.* 28, 638–640 (1970).
30. L. Heflinger, U.S. Patent 3600097 (17 August 1971).
31. J. Schwider, "Nonlinearities in Image Holography," *JOSA* 60(10), 1421 (1970).
32. J. Schwider, "Isophotes and Enhancement of Phase Sensitivity through Optical Filtering in Image Holography," *Materials of Third All-Union School on Holography*, Ul'yanovsk, USSR (1971).
33. A. Boutier (Ed.), *New Trends in Hypersonic Research*, NATO ASI Series, Series E: Applied Sciences, vol. 224 (1993).
34. J. Surget and G. Dunet, "Multi-Pass HI for the High Enthalpy Hypersonic Wind Tunnel F4," in [33], p.113–122.
35. F. Weigl, O. Friedrich, and A. Dougal, "Multi-Pass Non-Diffuse Holographic Interferometry," *IEEE J. Quantum Electronics* 6(1), 41–44 (1970).
36. J. Hsu and J. Trolinger, "Double-Plate Phase-Shifting Holographic Interferometry," *AIAA Paper* No. 93-2916 (1993).
37. W. Spring III, W. Yanta, K. Gross, and C. Lopez, "The Use of HI for Field Diagnostics," in [33], p. 97–112.
38. G. Dreiden, Y. Ostrovsky, and E. Shedova, "Holographic Interferometry in Simulated Raman Light," *Opt. Communic.* 4, 209–213 (1971).
39. Yu. Ostrovsky, M. Butusov, and G. Ostrovskaya, *Interferometry by Holography*, Springer-Verlag, New York (1980).

40. J. Trolinger, R. Hadson, B. Yip, and B. Batler, "Resonant HI—A Multipoint, Multiparameter Diagnostic Tool for Hypersonic Flow," in [33], p. 123–130.

41. I. Zeilikovich and A. Lyalikov, "Holographic Methods for Regulating the Sensitivity of Interference Measurements for Transparent Media Diagnostics," *Sov. Phys. Usp.* 34(1), 74–85 (1991).

42. J. Goodman, *Introduction to Fourier Optics,* McGraw-Hill, New York (1996).

43. G. Toker and J. Stricker, "Study of Suspended Vaporizing Volatile Liquid Droplets by an Enhanced Sensitivity Holographic Technique: Additional Results," *Int. J. Heat. Mass. Transfer* 41(16), 2553–2555 (1998).

Section II

44. C. Vest and D. Sweeney, "Holographic Interferometry with Both Beams Traversing the Object," *Applied Optics* 9(12), 2810–2812 (1970).

45. R. Anderson and J. Milton, "A Large Aperture Inexpensive Interferometer for Routine Flow Measurements," *ICIASF'89*, 394–399, 1989.

46. R. Smartt, "Special Applications of the Point-Diffraction Interferometer," in *Interferometry*, G. W. Hopkins ed., *Proc. SPIE* 192, 35 (1979).

47. L. Carr, M. Chandrasekhara, and S. Ahmed, "A Study of Dynamic Stall Using Real Time Interferometry," *AIAA* 91-0007 (1991).

48. M. Chandrasekhara, L. Carr, and M. Wilder, "Interferometric Investigations of Compressible Dynamic Stall over a Transiently Pitching Airfoil," *AIAA Jour.* 32(3), 586–593 (1994).

49. W. Howes, "Large-Aperture Interferometer with Local Reference Beam," *Applied Optics* 23(10), 1467–1473 (1984).

50. N. Spornik, "Holographic Interferometer with Diffraction Grating," *J. Applied Spectroscopy* 34(3), 534–536 (1981).

51. I. Kadushin and J. Rom, "Design of Intermittent Single Jack Flexible Nozzle Supersonic Wind Tunnel for Mach Number 1.5 to 4.0," *T.A.R. Report No. 86*, Technion (1968).

52. R. Small, V. Sernas, and R. Page, "Single Beam Schlieren Interferometer Using a Wollaston Prism," *Applied Optics* 11, 1332–1333 (1972).

53. W. Merzkirch, "Generalized Analysis of Shearing Interferometers As Applied for Gas Dynamic Studies," *Applied Optics* 13, 409–413 (1974).

54. J. S. Wyant, "Double Frequency Grating Lateral Shear Interferometer," *Applied Optics* 12(9), 2057–2060 (1973).

55. O. Bringdahl, "Shearing Interferometry by Wave Front Reconstruction," *JOSA* 58(7), 865–871 (1968).

56. O. Bringhdal, "Longitudinally Reversed Shearing Interferometry" *JOSA* 59(2), 142–146 (1969).

57. G. Toker, D. Levin, and J. Stricker, "Dual Hologram Shearing Interference Technique for Wind Tunnel Flow Field Testing," *Experiments in Fluids* 23(4), 341–346 (1997).

58. A. Jones, M. Schwar, and F. Weinberg, "Generalized Variable Shear Interferometry for the Study of Stationary and Moving Refractive Index Fields with the Use of Laser Light," *Proc. Roy. Soc. Lond. A* 322, 119–135 (1971).

59. G. Toker, D. Levin, and A. Lessin, "Investigation of Supersonic Flows by Dual Hologram Shearing Interferometry," *Proceedings of the Third International Conference on Experimental Fluid Mechanics, p. 27*, Korolev, Russia (1997).

60. A. Lessin, D. Levin, V. Suponitsky, and G. Toker, "Direct Measurements of Aero-Optical Effects in the Technion Supersonic Wind Tunnel," *Proc. SPIE* 3432, 57 (1998).

61. R. Goldstein (Ed.), *Fluid Mechanics Measurements*, University of Minnesota, Hemisphere Publishing Corp. (1983).

62. W. Merzkirch, *Flow Visualization*, Academic Press, London (1987).

63. R. Ladenburg (Ed.), *Physical Measurements in Gas Dynamic and Combustion*, Princeton University Press, Princeton, NJ (1954).

64. J. Stricker and B. Zakharin, "3-D Turbulent Density Field Diagnostics by Tomographic Moiré Technique," *Experiments in Fluids* 23(1), 76–85 (1997).

65. I. Levy and B. Golovanevsky, "Atomization Characteristics of n-Heptone and n-Hexane Droplets," *Proceedings of 2nd Int. Conf. on Exp. Fl. Mech. (ICEEFM 94)*, 118–125, Torino, Italy (1994).

66. G. Toker, D. Levin, and A. Lessin, "Dual Hologram Shearing Interference Technique with Enhanced Sensitivity for Wind Tunnel Testing," *Experiments in Fluids* 25(5/6), 519–521 (1998).

67. G. Toker and D. Levin, "Dual Hologram Shearing Interferometry with Regulated Sensitivity," *Applied Optics* 37(8), 5162–5168 (1998).

68. A. Lyalikov, "Enhancement of Sensitivity of Measurements of the Holographic Later-Shear Interferometry in Studies of Fast Processes," *Optics and Spectroscopy* 93(3), 472–476 (2002).

69. A. Lyalikov, "Extension of the Controlled Sensitivity Range in Holographic Interferometry with a Lateral-Shear," *Optics and Spectroscopy* 93(3), 477–481 (2002).

70. S. Gribin and G. Ostrovskaya, "Problems of Interpretation of Holographic Interferograms Near Shock Wave Front," *Technical Physics* 43(9), 1087–1090 (1998).

71. F. Weigl, "A Generalized Technique of Two-Wavelength Non-Diffuse Holographic Interferometry," *Appl. Opt.* 10, 187–192 (1971).

72. F. Weigl, "Two-Wavelength Holographic Interferometry for Transparent Media Using a Diffraction Grating," *Appl. Opt.* 10, 1083–1086 (1971).

73. K. Mustafin and V. Seleznev, "Holographic Interferometry with Variable Sensitivity," *Opt. Spectrosc* 32, 532–535 (1972).

74. G. Toker, D. Levin, and A. Lessin, "Phase Measurements with Reduced Sensitivity of a Supersonic Shock Flow by Two Wavelength Dual Hologram Shearing Interference Technique," *International Congress on Instrumentation in Aerospace Simulation Facilities*, 313–321 (1997).

75. R. Bracewell, *The Fourier Transform and Its Applications*, McGraw-Hill, New York (1970).

76. G. Toker and D. Levin, "Real-Time Holographic Shearing Interference Technique for Wind Tunnel Testing," *Proceeding of 8th International Symposium On Flow Visualization*, #241, Naples, Italy (1998).

77. G. Toker and N. Korneev, "Disappearance of holographic and Interference Fringes Accompanies Optical Diagnostics of a Supersonic Bow Shock Flow," *Optik* 119, 112–116 (2008).

78. B. Zakharin and J. Stricker, "Fourier Optics Analysis of Schlieren Images," Ninth International Symposium on Flow Visualization, Edinburgh, UK, August 22–25 (2000).

79. R. Yano, V. Contini, E. Pl-Ogravenjes, P. Palm, S. Merriman, S. I. Aithal, and I. Adamovich, "Supersonic Nonequilibrium Plasma Wind-Tunnel Measurements of Shock Modification and Flow Visualization," *AIAA J.* 38(10), 1879–1888 (2000).

80. T. McIntyre, A. Bishop, A. Thomas, M. Wegener, and H. Rubinsztein-Dunlop, "Emission and Holographic Interferometry Measurements in a Superorbital Expansion Tube," *AIAA J.* 36(6), 1049–1054 (1998).

81. R. Small, V. Sernas, and R. Page, "Single Beam Schlieren Interferometer Using a Wollaston Prism," *Appl. Opt.* 11(4), 858–862 (1972).

82. G. Toker and D. Levin, "Dual Hologram Shearing Interferometry with Regulated Sensitivity," *Appl. Opt.* 37(8), 5162–5168 (1998).

83. M. Born and E. Wolf, *Principles of Optics*, 6th ed., Pergamon Press, New York (1989).

84. V. Yassinsky, "Correction of the Results of the Interferometric Measurements by Using the Speckle Patterns," in *Proceedings of the Seventh International Conference on Optics within Life Sciences*, Luzern, Switzerland (2002).

85. Yu. I. Ostrovsky, *Holography and Its Applications*, Mir Publications, Moscow (1973).

86. A. K. Beketova and A. F. Belozerov, *Holographic Interferometry of Phase Objects*, Nauka Publishing, L., Moscow (1979).

Section III

87. T. Kreis, *Handbook on Holographic Interferometry: Optical and Digital Methods*, Wiley-VCH Verlaag GmbH & Co. KGaA, Weinheim, Germany (2005).

88. U. Schnars and W. Jüptner, *Digital Holography: Digital Hologram Recording, Numerical Reconstruction and Related Techniques*, Springer-Verlag, Berlin Heidelberg (2005).

89. U. Schnars, "Direct Phase Determination in Hologram Interferometry with Use of Digitally Recorded Holograms," *J. Opt. Soc. Am. A* 11(7), 2011 (1994).

90. U. Schnars, P. Werner, and W. Juptner, "Digital Recording and Numerical Reconstruction of Holograms," *Meas. Sci. Technology* 13, R85 (2002).

91. T. M. Kreis, "Frequency Analysis of Digital Holography," *Opt. Eng.* 41, 771 (2002).

92. M. Takeda, H. Ina, and S. Kobayashi, "Fourier-Transform Method of Fringe-Pattern Analysis for Computer-Based Topography and Interferometry," *J. Opt. Soc. Am.* 72(1), 156–160 (1982).

93. W. W. Macy, Jr., "Two-Dimensional Fringe-Pattern Analysis," *Applied Optics* 22(23), 3898 (1983).

94. S. Toyooka and M. Tominaga, "Spatial Fringe Scanning for Optical Phase Measurements," *Optics Communications*, V. 51(2), 68–70 (1984).

95. K. A. Nugent, "Interferogram Analysis Using an Accurate Fully Automatic Algorithm," *Applied Optics* 24(18), 3101 (1985).

96. T. Kreis, "Digital Holographic Interference-Phase Measurement Using the Fourier-Transform Method," *J. Opt. Soc. Am. A* 3(6), 847 (1986).

97. R. J. Green and J. G. Walker, "Investigation of the Fourier-Transform Method of Fringe Pattern Analysis," *Optics and Lasers in Engineering* 8, 29–44 (1988).

98. Jozwicka, A., Kujawinska, M., "Digital Holographic Tomograph for Amplitude-Phase Microelements Testing," *Lasers and Electro-Optics Europe,2005.* CLEO/Europe. p. 459 (2005).

99. O. Bringdahl, *Applications of Shearing Interferometry, Progress in Optics*, v. 4, p. 39–83, Amsterdam: North-Holland (1965).

100. E. Waetzmann, "Interferenzmethode zur untersuchung der abbildungsfehler optischer systeme," *Ann. Phys.* 39, 1042–1052 (1912).

101. W. J. Bates, "A Wave Front Shearing Interferometer," *Proc. Phys. Soc.* 59, 940–950 (1947).

102. R. L. Drew, "A Simplified Shearing Interferometer" *Proc. Phys. Soc. B* 64, 1005 (1951).

103. H. Oertel, *Optische Stromungsmesstechnik*, p. 337–364, G. Braun, Karlsruhe, Germany (1989).

104. N. Qi et al., "Z Pinch Imploding Plasma Density Profile Measurements Using a Two-Frame Laser Shearing Interferometer," *Plasma Science, IEEE Transactions on* 30(1), 227–238 (2002).

105. N. Ramesh and W. Merzkirch, "Combined Convective and Radiative Heat Transfer in Side-Vented Open Cavities," *International Journal of Heat and Fluid Flow* 22(2), 180–187 (2001).

106. G. Pletzier, H. Jager, and T. Neger, "High-Accuracy Differential Interferometry for the Investigation of Phase Objects," *Meas. Sci. Technol.* 4, 649–658 (1993).

107. U. Schnars and W. Juptner, "Digital Recording and Reconstruction of Holograms in Hologram Interferometry and Shearography," *Applied Optics* 33(20), 4373–4377 (1994).

108. P. Ferraro, D. Alfieri, S. De Nicola, L. De Petrocellis, A. Finizio, and G. Pierattini, "Quantitative Phase-Contrast Microscopy by a Lateral Shear Approach to Digital Holographic Image Reconstruction," *Optics Letters* 31(10), 1405–1407 (2006).

109. B. Kemper and G. von Bally, "Digital Holographic Microscopy for Live Cell Applications and Technical Inspection," *Applied Optics* 47(4), A52–A61 (2008).

110. G. Haijun and T. Zhiwei, "Phase Information Reconstruction Based on Lateral Shearing Approach in Digital Holography," *29th Chinese Control Conf.*, p. 3056–3058 (2010).

111. I. Gökalp, C. Chauveau, and O. Simon, "Mass Transfer from Liquid Droplets in Turbulent Flow," *Combustion and Flame* 89, 286 (1992).

112. M. Birouk, C. Chauveau, and I. Gökalp, "Turbulent Effects on the Vaporization of Monocomponent Single Droplets," *Combust. Sci. Technology* 113–114, 413–428 (1996).

113. L. Vasil'ev, *Schlieren methods*, Keter, New York (1971).

114. E. Eckert and R. Goldstaein, *Measurements in Heat Transfer,* 2nd ed. McGraw Hill, New York (1976).

115. D. Kastel, K. Kihm, and L. Flechter, "Study of Laminar Thermal Boundary Layer Occurring Around the Leading Edge of a Vertical Isothermal Wall Using Specklegram Technology," *Experiments in Fluids* 13, 249–256 (1992).

116. G. Settles, "Color-Coding Schlieren Techniques for the Optical Study for Heat and Fluid Flow," *Int. J. Heat Fluid Flow* 6, 3–15 (1985).

117. G. Toker and D. Levin," Comprehensive Optical Diagnostics of Complex Flow Fields," *SPIE Conference #3110-81*, 574–585 (1997).

118. G. Toker, D. Levin, J. Stricker, and A. Lessin, "Optical Diagnostics of Complex Flow Field in a Supersonic Wind Tunnel," *Proceed. of the 37 Israel Ann. Conf. on Aerospace Sciences*, 86–96 (1997).

119. J. Trolinger, "Flow Visualization Holography," *Opt. Eng.* 14(5), 470–481 (1975).

120. D. Kastel and G. Eitelberg, "A Combined Holographic Interferometer and Laser-Schlieren System applied to High Temperature, High Velocity Flows," *ISIASF'95*, 12.1–12.7 (1995).

121. V. Oskay, "Interferometry with Hypersonic Free-Flight Models," *AIAA Journal* 5(1), 156–157 (1967).

122. G. Settles, *Schlieren & Shadowgraph Techniques*, Springer Verlag, New York (2001).

123. B. Zakharin, J. Stricker, and G. Toker, "Laser-Induced Spark Schlieren Imaging," *AIAA Journal* 37(9), 1133–1135 (1999).

124. J. Stricker, E. Keren, and O. Kafri, "Axisymmetric Density Field Measurements by Moire Deflectometry," *AIAA Journal* 21(12), 1767–1769 (1983).

125. O. Kafri and I. Glatt, *The Physics of Moire Metrology*, Wiley Series in Pure and Applied Optics, ser. ed. J. W. Goodman, John Wiley and Sons, Inc., New York (1990).

126. J. Stricker, "Analysis of 3-D Phase Objects by Moiré Deflectometry," *Applied Optics* 23(20), 3657–3659 (1984).

127. J. Stricker and J. Politch, "Holographic Moiré Deflectometry—A Method for Stiff Density Field Analysis," *Appl. Phys. Lett.* 44(8), 723–725 (1984).

128. D. P. Yan, A. Z. He, and X. W. Ni, "New Method of Asymmetric Flow Field Measurement in Hypersonic Shock Tunnel," *Applied Optics* 30(7), 770–774 (1991).

129. E. Harvey, M. Bouchard, and P. Langois, "Holographic Interferometry and Moiré Deflectometry for Visualization and Analysis of Low-Gravity Experiments on Laser Materials Processing," *Optical Engineering* 32(9), 2143–2154 (1993).

130. M. Born and E. Wolf, *Principles of Optics*, 2nd ed., Pergamon Press, New York (1964).

131. R. Pitlak and R. Page, "Pulsed Lasers for Holographic Interferometry," *Optical Engineering* 24(4), 639–644 (1985).

132. N. V. Karlov, *Lectures on Quantum Electronics*, CRC Press, Boca Raton, FL (1993).

Index